GEOSPATIAL ANALYSIS USING OPEN-SOURCE SOFTWARE

The Editors

Dr. Vinod Kumar Yadav is Senior Scientist (Agricultural Statistics) at Fisheries Economics, Extension and Statistics (FEES) Division, ICAR-Central Institute of Fisheries Education (CIFE) Mumbai. He has more than 15 years of teaching and research Experience. His research interest includes Modelling and Forecasting, GIS and Remote Sensing, Climate change, Artificial Intelligence and Big Data Analytics. He is teaching several courses at Master and Ph.D. level like Mathematics, Statistics Methods. Advanced statistical Methods, GIS and Remote Sensing, Artificial Intelligence and Big data Analytics. Dr. Yadav received Master degree from Indian Agricultural Research Institute (IARI) Pusa New Delhi and Ph.D. from IIT Bombay.

Dr. Vidya Shree Bharti completed her MSc (Soil Science) from G.B. Pant University of Science and Technology, Uttrakhand and Ph.D. from Indian Institute of Technology, Mumbai, India. She joined her Agricultural Research Service as a Scientist in the Indian Council of Agricultural Research in 2005. She served as a Scientist for two years at the Indian Institute of Soil Science, Bhopal, and in the year 2007 she got transferred to the Central Institute of Fisheries Education, Mumbai. She published articles, book chapters, training manuals, and the papers in National and International journals. Since 2012 She is working as a senior scientist in the Aquatic Environment and Health Management division of ICAR- CIFE Mumbai.

Dr. Arpita Sharma is Principal Scientist and Head of Fisheries Economics, Extension and Statistics Division/Social Sciences Division at Indian Council of Agricultural Research-Central Institute of Fisheries Education, a Deemed University, located in Mumbai, India.

GEOSPATIAL ANALYSIS USING OPEN-SOURCE SOFTWARE

Dr. Vinod Kumar Yadav

Dr. Vidya Shree Bharti

Dr. Arpita Sharma

2026

Daya Publishing House®

A Division of

Astral International Pvt. Ltd.

New Delhi – 110 002

ISBN: 978-93-5919-659-6 (HB)

Published by : **Daya Publishing House®**
A Division of
Astral International Pvt. Ltd.
– ISO 9001:2015 Certified Company –
4736/23, Ansari Road, Darya Ganj
New Delhi-110 002
Ph. 011-43549197, 8130496929
E-mail: info@astralint.com
Website: www.astralint.com

Printed at : **Replika Press Pvt. Ltd.**

ACKNOWLEDGEMENT

The author is extremely thankful to Dr. Ravishankar C. N., Director/Vice-Chancellor, ICAR-CIFE, for providing all necessary facilities to carry out the work. The author expresses gratitude to Dr. Arpita Sharma , Head FEES(Fisheries Economics, Extension and Statistics) Division, CIFE Mumbai for her constant motivation and encouragement in every step to accomplish this journey. The author expresses his pleasure and thanks to all his family members who always stand on his side for all the support and encouragement provided to accomplish this journey. Finally, author would like to thank his colleagues at FEESD, his students at ICAR-CIFE and whoever in a wide range of spectrum helped him. Special thanks are to Mr. Liton Paul , Ms. Chandani Dave, Mr. Mahesh Sharma, his M.F.Sc. (FEC) students for their painstaking efforts in evaluating the manual and providing useful suggestions.

Dr. Vinod Kumar Yadav

FOREWORD

ICAR-Central Institute of Fisheries Education is playing a pivotal role in human resource development in all the major disciplines of Fisheries Science. Apart from conducting post graduate programs, the faculty members are involved in imparting training in the emerging areas of fisheries and aquaculture. Remote Sensing and GIS is one such area that has got tremendous potential for its application in fisheries and aquaculture sector such as soil and water quality mapping, disease outbreak mapping, mangrove mapping, identifying potential fishing zones etc. CIFE is a pioneer institute to take up research in this field and has established the working facilities for teaching, research and training at the Fisheries Economics, Extension & Statistics Division.

Book on "Geospatial Analysis Using Open-Source Software" is a well-designed content that covers different aspects of GIS and remote sensing technology and is useful to various stakeholders of aquaculture and fisheries sector. The book prepared for this programme is a compilation of different chapters and important relevant protocols. I am sure that this would serve as a useful reference material for the research scholars, fisheries officers, progressive farmers, entrepreneurs and other fisheries professionals who are interested in GIS and remote sensing technology for fisheries and aquaculture.

Ravishankar C. N.
Director/Vice-Chancellor,
ICAR-Central Institute of Fisheries Education (CIFE) Mumbai

PREFACE

Knowledge in the field of Remote Sensing and GIS is advancing at a rapid pace leading to new and varied applications. Remote Sensing and GIS tool is now used by people working in every area of science and management. Several applications of these technologies are possible in fisheries and aquaculture sector too so the fisheries development officers, entrepreneurs, farmers and fisheries professional need to be equipped with this knowledge.

Realizing the importance of this subject, a book on "Geo-spatial Analytics using Open-Source Software" were written by Authors(Dr. Vinod Kumar Yadav, Dr. Vidya S. Bharti and Dr. Arpita Sharma) of ICAR-Central Institute of Fisheries Education (ICAR- CIFE), Mumbai.

This book was conceptualized for research scholars, fisheries development officers, entrepreneurs, technical experts and fish farmers who have interest in Remote Sensing and GIS technology. The pedagogy of this book involved theory and practical section to provide comprehensive learning and thus provides complete theoretical and practical knowledge about retrieval of satellite image information, digital image processing, potential fishing zone assessment and mapping through GIS and remote sensing technology.

QGIS software is selected for practical exercises. It is a free and open-source desktop platform. It allows users to analyze and edit spatial information, in addition to composing and exporting graphical maps. QGIS supports raster, vector and mesh layers, also processes different image processing tasks by using different plugins. Experts contributing to the chapters of this book were drawn from various disciplines and institutes. The book chapters and protocols provided by them have been compiled and edited to serve as a valuable reference material for readers.

We would like to place on record our sincere gratitude to Dr. Ravishankar C. N., Director, ICAR- CIFE for his unstinted support and motivation in organizing this unique training programme.

Vinod Kumar Yadav

Vidya Shree Bharti

Arpita Sharma

CONTENTS

CHAPTER 1

INTRODUCTION TO GEOSPATIAL TECHNOLOGY AND IT'S APPLICATIONS

Vinod Kumar Yadav, Ranjana Choudhury and Mahesh Sharma

ICAR-Central Institute of fisheries Education, Mumbai-400061

Introduction

The number of satellite missions dedicated to remote sensing (or Earth Observation (EO)) has increased significantly over the past decade and will further increase over the coming decade and beyond. Data from these missions offer the potential for contributing to the security of human existence on Earth in a wealth of different ways. Although there have been many demonstrations of the value of EO satellites to development issues such as food production, resource management and environment characterization, it is recognized that there is a role for transferring that knowledge to developing countries, and it is hoped thatthis publication will stimulate, encourage and assist countries inexperienced in the use of satellite EO data to acquire the necessary expertise to exploit and realize its full potential. The raw data from satellite EO often requires complex processing both to correct for atmospheric and geometric distortions and to derive information from the data and imagery. Additional data, such as that from in-situ sources, is sometimes used to supplement the EO data in this processing. Such data is particularly useful for calibrating or validating models.

Definition of Remote Sensing

The science of acquiring information about the earth using instruments which are remote to the earth's surface, usually from aircraft or satellites. Instruments may use visi`ble light, infrared or radar to obtain data. Remote sensing offers the ability to observe and collect data for large areas relatively quickly, and is an important source of data for GIS. Also, it is defined as the transport of information from an object to a receiver (observer) by means of radiation transmitted through the atmosphere. The interaction between the radiation and the object of interest conveys information required on the nature of the object (eg. reflection coefficient, emittance, roughness).

Examples

i. The reflection of sunlight from vegetation will give information on the reflection coefficient of the object and its spectral variation, and thus on the nature of the object (green trees, etc.).

ii. Microwave radiation transmitted from a radar system and scattered from a rain cloud in the back direction to a receiver will give information on the raindrop size and intensity.

Passive and active sensing

The first example above is an example of passive remote sensing, where the reflected radiation observed originates from a natural source - the sun. The second example is an example of active remote sensing, where the scattered radiation originates from a specially designed active radar system.

Electromagnetic Radiation

Radiation can be observed either as a wave motion, or as single discrete packets of energy, photons. The two descriptions are not really contradictory. The energy is emitted as photons, but its statistical distribution over time is described by a wave.

Normally, one is dealing with a large number of photons arriving in a short time, and the radiation can be treated physically as a wave motion. However, in the visible and ultraviolet regions, very weak sources are typified by the detection of single photons. The wave theory of radiation has been developed extensively. It impacts on remote sensing in the way that radiation is reflected at a surface and transmitted, absorbed and scattered in a medium.

The energy E of a photon is given by:

$$E = h\nu \quad (1)$$

$$= hc/\lambda \quad (2)$$

where c, ν and λ are the velocity, frequency and wavelength of the radiation respectively, and h is Planck's constant.

The Electromagnetic Spectrum

Electromagnetic radiation covers a very large range of wavelengths. In Remote Sensing we are concerned with radiation from the ultraviolet (UV), which has wavelengths of from 0.3 to 0.4 um (10^-6 m) to radar wavelengths in the region of 10 cm (10^-l m) (see Figure below). Thus, the phenomena observed in the various wavelength regions differ considerably.

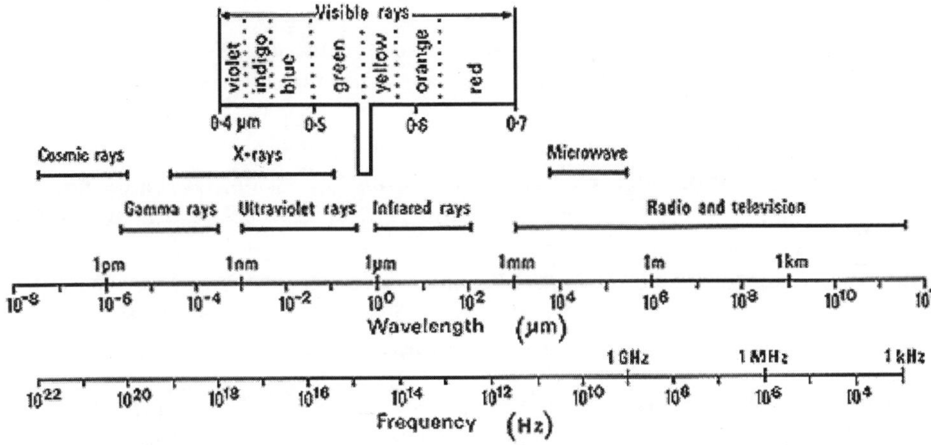

Fig. 1.1: Range of electromagnetic wavelengths and the transmission through the atmosphere

Reflection from Vegetation

Healthy, growing vegetation appears green because there is selective absorption in chlorophyll bands outside the green wavelengths. The absorption is only moderate so thatthe green light is reflected and scattered at the cellular boundaries to appear green both inreflection and transmission. Because of multiple reflections emergent natural light is non-polarized. At wavelengths beyond about 0.65 us the reflection becomes strong.

Data Resolutions

The resolution of an image refers to the potential detail provided by the imagery. In remote sensing we refer to four types of resolution: spatial, spectral, radiometric and temporal.

Spatial resolution

Spatial Resolution refers to the size of the smallest feature that can be detected by a satellitesensor or displayed in a satellite image. It is usually presented as a single value representingthe length of one side of a square. For example, a spatial resolution of 250m means that one pixel represents an area 250 by 250 meters on the ground.

Effect of Spatial Resolution

- When resolution is very high, we perceive individual objects such as buildings or Roads
- When resolution is medium, we perceive very large objects as individual features, andareas as textured regions
- When resolution is coarse, we perceive color or tone variations, and large area-basedfeatures.
- High resolution images are rich in spatial information, Multispectral information and Textural information.

Spectral Resolution

Spectral Resolution refers to the ability of a satellite sensor to measure specific wavlengths of the electromagnetic spectrum. The finer the spectral resolution, the narrower the wavelength ranges for a particular channel or band. This is the ability of the sensor to distinguish differences in reflectance of ground features in different wavelengths. This is Essential to discriminate between subclass of a broad class such as vegetation, road, water body etc.

Radiometric resolution

Radiometric Resolution: The radiometric resolution of an imaging system describes its ability to discriminate very slight differences in energy. The finer the radiometric resolution of a sensor the more sensitive it is to detecting small differences in reflected or emitted energy.

Temporal resolution

Temporal resolution refers to the time between images. The capability for satellites to provideimages of the same geographical area more frequently has increased dramatically since the dawn of the space age.

Relevance of Remotely Sensed Images Remote Sensing provides vital data for manycritical applications

- Resources management
- Environmental monitoring
- Defense
- Urban / rural development and planning
- Crop yield forecasting
- Hazard zonation and disaster mitigation

Strength of Satellite Remote Sensing

- Large aerial coverage since viewed from a high altitude
- Repetitive coverage with earth orbiting satellites

- Observation in wide range of wavelengths sensitive to reflectance/ emission characteristics of terrain features including and beyond the visible spectrum

Digital Image Processing

Processing procedures fall into three broad categories: Image Restoration (Preprocessing); Image Enhancement; and Classification and Information Extraction. Good contrast and brightness essential for visually appreciating the content. Image data are modified to improve or enhance the visual quality. Under Information Extraction, ratioing and principal components analysis have elements of Enhancement but lead to images that can be interpreted directly for recognition and identification of classes and features. Also included in the third category but treated outside this lecture is Change Detection and Pattern recognition. The data in satellite remote sensing is in the form of Digital Number or DN. It is said that the radiances, such as reflectance and emittances, which vary through a continuous range of values are digitized onboard the spacecraft after initially being measured by the sensor(s) in use. Ground instrument data can also be digitized at the timeof collection. Or, imagery obtained by conventional photography is capable of digitization. A DN is simply one of a set of numbers based on powers of 2, such as 2^6 or 64. Resampling is one approach commonly used to produce better estimates of the DN valuesfor individual pixels. After the various geometric corrections and translations have been applied, the net effect is that the resulting redistribution of pixels involves their spatial displacements to new, more accurate relative positions

Classification

This is the process of classifying multispectral images into patterns of varying gray or assigned colors that represent either clusters of statistically different sets of multiband data (radiances expressed by their DN values), some of which can be correlated with separable classes/features/materials (Unsupervised Classification), or numerical discriminators composed of these sets of data that have been grouped and specified by associating each with a particular class, etc. whose identity is known independently and which has representative areas (training sites) within the image where that class is located (SupervisedClassification).

Remote sensed data product

- Major geophysical data products derived from satellite remote sensing techniques that are commonly used in the ecosystem studies and also to familiarize a major data provider named 'Asia Pacific Data Research Centre (APDRC).
- There are different websites which provide various geophysical data products and that are offered either freely accessible or payment basis. Here we introduce three websites named as Ocean Colour web, GIOVANNI

both are being maintained by NASA and Asia Pacific Data Research Centre (APDRC) website by International Pacific Research Centre (IPRC). GiOVANNl is an acronym for the Goddard Earth Sciences Data and Information Services Centre (GES-DISC) Interactive Online Visualization and Analysis Infrastructure. The first two are specially dedicated for ocean colour data products and the latter provides all available data products.

Geographic Information System (GIS)

The GIS field is characterised by a great diversity of applications and concepts developed in many areas- agriculture, statistics, computer science, graphics, mathematics, surveying, cartography, geology, geography, database technology, resource management and decision making etc. For the purpose of understanding, the following definition of GIS encompasses most of the concepts.

- A Geographical Information System is a tool, which can accept large volumes of spatial data derived from a variety of sources, retrieve, manipulate, analyse and display them according to user-defined specifications.
- An information technology which stores, analyses, and displays both spatial and non-spatial data
- A powerful set of tools collecting, storing, retrieving at will, transforming and displaying spatial data from the real world

Data in GIS

Broadly, the basic data for any GIS application can be categorized as;

a. ***Spatial Data*** consisting of maps which have been prepared either with the help of field surveys or with the help of interpreted remotely sensed data (RS). Remote sensing data is a classic source of data on natural resources for a region and provides a record of the continuum of resource status because of its repetitive coverage. Remotely sensed data in the form of satellite imageries can be used to study and monitor land features, natural resources and dynamic aspects of human activities for preparation of thematic maps.

b. ***Non-Spatial Data*** is attributes as complimentary to the spatial data and describe what is at a point, along a line or in a polygon and as a socio-economic characteristic fromcensus or other sources. The attributes of a soil category could be the depth of soil, texture, erosion, drainage etc. and for geological category could be the rock type, its age,major composition etc. The socio- economic characteristics could be the demographic data, occupation data for a village etc.

The GIS will have to be the workhorse of integrated database system as both spatial and non-spatial data needs to be handled. The GIS package offers efficient utilities to handleboth these data sets and also allows for the spatial database organization along with non- spatial database organization. It is also capable to transform as well as integrate these two different kinds of information.

Spatial data-describe the location of spatial features which may be discrete or continuous. The features shown on a map, e.g., lakes, buildings and contours can be thought of as discrete objects or features. Discrete features include points(wells), lines (road) and arc(land use types). Continuous features are features that exist spatially between observations. Some characteristics exist everywhere and vary continuously over the earth's surface. For example,elevation, atmospheric temperature and pressure, natural vegetation or soil type does not present any visible boundary in the real world.

Attributes are either the qualitative characteristics of the spatial data or descriptive information about geographical features. Attributes are stored in form of tables where each column of the table describes one attribute and each row of the table corresponds to a feature.

Data Structure

In order to represent the spatial information along their attributes a data model, which is set of logical definitions or rules for characterizing the geographical data, is adopted. The data model represents the linkage between the real-world domain of geographical data and computer or GIS presentations of these features. Different type of structures has been usedas far as GIS is concerned. They are Raster model, Vector model, Quadtree model etc. The first two are most popular in GIS packages available in the market.

Raster Model represents the image with help of square lattice grids. In this case system stores an image by assigning a series of values (generally an integer ranging between 0 to 255) to each cell identified by its Cartesian co-ordinates in space.

Vector Model represents the geographical feature by a set of co-ordinates vectors as xy- coordinates define points, lines and polygons. The basic premise of the vector-based structuring is to define a two-dimensional space where features are represented by coordinates on the two axes.

The vector image is more pleasant to the eye and more accurate representation to the reality. This implies that a vector-based GIS is a better way to produce graphical output. On the other hand, vector image is more complex and requires more advanced technology in terms of hardware. Furthermore, from analysis point of view data in vector format presents more problems if we want to apply complex spatial statistical procedures. In contrasts, due to its regularity, a raster image is more easily accessible and is the natural format for many spatial techniques. Furthermore, it requires a less advanced technology. On the other hand, raster-based image produces less pleasant graphic outputs. Further a raster image is based on quantization of reality and as such it can lead to serious estimation errors when we are interested in geometric and topologic characteristics like area or perimeters. The choice between vector and raster is crucial one and depends on specific aim for which GIS is designed.

A GIS has four main functional subsystems. These are:

- a data input subsystem;
- a data storage and retrieval subsystem;
- a data manipulation and analysis subsystem; and
- a data output and display subsystem.

A **data input** subsystem allows the user to capture, collect, and transform spatial and thematic data into digital form. The data inputs are usually derived from a combination of hard copy maps, aerial photographs, remotely sensed images, reports, survey documents, etc. The **data storage** and retrieval subsystem organizes the data, spatial and attribute, in a form which permits it to be quickly retrieved by the user for analysis, and permits rapid and accurate updates to be made to the database. This component usually involves use of a database management system (DBMS) for maintaining attribute data. Spatial data is usually encoded and maintained in a proprietary file format.

The **data manipulation and analysis** subsystem allows the user to define and execute spatial and attribute procedures to generate derived information. This subsystem is commonly thought of as the heart of a GIS.

The data output subsystem allows the user to generate graphic displays, normally maps, and tabular reports representing derived information products.

Integrated Modelling in GIS

Integration, in a GIS context, is the synthesis of spatial and non-spatial information with in the frame work of an application. By performing the operations across the two sets of information in tandem, a far richer set of questions can be answered and a far broader range of problems can be solved than in a system that handle just attribute or spatial data alone. All problems of GIS based integration involve a conjuctive analysis of multi-parameter data.The multi-parameter data include different spatial inputs say, maps of land use, soils, slopes,terrain etc. and other non-spatial data sets. The GIS allows for the integration of these data sets so as to obtain a composite information set. However, the important aspect is the interpretation and the analysis of the integrated information sets. Toward this there are two aspects that are important for integrated model building, first is the criterion that defines the logic for the analysis of the composite information set and second is the relative importance or weightage of each of the parameter for the end objectives.

Applications of Remote Sensing-GIS in Various fields

1. Environmental Science

- Environmental Impact Analysis (EIA)
  - » EIA is an important policy initiative to conserve natural resources and environment.
  - » Many human activities produce potential adverse environmental effects which include the construction and operation of highways, rail roads, pipelines, airports, radioactive waste disposal and more.
  - » Environmental impact statements are usually required to contain specific information on the magnitude and characteristics of environmental impact.
  - » The EIA can be carried out efficiently by the help of GIS, by integrating various GIS layers, assessment of natural features can be performed.
- Disaster Management
  - » GIS can help with risk management and analysis by displaying which areas spatially are likely to be prone to natural or man-made disasters. When such disasters are identified, preventive measures can be developed.
  - » Zoning of Landslides hazard: - Hazard zonation is the process of ranking
  - » different parts of an area according to the degrees of actual or potential hazard from landslides. The evaluation of landslide hazard has been complex task andby use of GIS it has become possible to efficiently collect, manipulate and integrate a variety of spatial data such as geological, structural, surface cover and slope characteristics of an area, which can be used for hazard zonation.
  - » Estimation of flood damage: - GIS helps to document the need for federal disaster relief funds, when appropriate and can be utilized by insurance agencies to assist in assessing monetary value of property loss. A local government need to map flooding risk areas for evaluate the flood potential level in the surrounding area. The damage can be well estimate and can be shown using digital maps.
  - » Identification of Volcanic Hazard: - Volcanic hazard to human life and environment include hot avalanches, hot particles gas clouds, lava flows and flooding. Potential volcanic hazard zone can be recognized by the characteristic historical records of volcanic activities, it can incorporate with GIS, thus an impact assessment study on volcanic hazards deals with economic loss and loss of lives and property in densely populated areas.

- Determination of land cover and land use
  - » Land cover means the feature that is covering the barren surface.
  - » Land use means the area in the surface utilized for particular use.
  - » The role of GIS technology in land use and land cover applications is that wecan determine land use/land cover changes in the different areas.
  - » Also, it can detect and estimate the changes in the land use/ land coverpattern within time.
  - » It enables to find out sudden changes in land use and land cover either bynatural forces or by other Anthropogenic activities like deforestation.
- Management of Natural Resources
  - » By the help of GIS technology, the agricultural, water and forest resources can be well maintained and manage. Foresters can easily monitor forest condition. GIS is also used in afforestation planning.
  - » Agricultural management includes managing crop yield, monitoring crop rotation, Irrigation status and more.
  - » Water is one of the most essential constituents of the environment. GIS is usedto analyze geographic distribution of water resources.
  - » Interrelationship between natural resources can also be studied, i.e. forest cover reducing the storm water runoff and tree canopy storing approximately 215,000 tons carbon.
  - » GIS is also used in afforestation planning.
  - » Soil Mapping: - Soil mapping provides resource information about an area/site. It helps in understanding soil suitability for various land use activities. It is essential for preventing environmental deterioration associated with misuse of land. GIS Helps to identify soil types in an area and to delineate soil boundaries. It is used for the identification and classification of soil. Soil map is widely used by the farmers in developed countries to retain soil nutrients and earn maximum yield.
  - » Wetland Mapping: - Wetlands contribute to a healthy environment and retain water during dry periods, thus keeping the water table high and relatively stable. During the flooding they act to reduce flood levels and to trap suspended solids and attached nutrients. GIS provide options for wetland mapping and design projects for wetland conservation quickly with the help of GIS. Integration with Remote Sensing data helps to complete wetland mapping on various scale. We can create a wetland digital data bank with species information using GIS.

2. Ecology

Remote sensing has been used for mapping the distribution of forest ecosystems, global fluctuations in plant productivity with season, and the three-dimensional (3D) structure of forests. Remote sensing has been used in a diverse range of forest ecology and management applications from mapping invasive species to monitoring land-cover changes, such as habitat fragmentation, to estimating biophysical and biochemical properties of forests. Remote sensing allows monitoring of important ecological variables, such as land-cover change, measurement of Vegetation cover, structure, vegetation chemistry and moisture, biodiversity, and soil characteristics. These variables are critical for understanding forest ecosystem functions and processes, as well as classifying forests into specific communities, ecosystems, and biomes. Historical remote-sensing data allow us to go back in time to look at the causes of present environmental issues.

3. Social/Political Sciences

- Public health: Public health seeks to address the health issues affecting people at the basic level or even at a larger national perspective. In the case of disease outbreaks, GIS is among the few systems that the government uses to do statistics. GIS is applied to ensure prevention and control of certain diseases. Its ability to provide the mapping capabilities has proved to work well with the health sector. E.g. COVID-19 Patterns of past diseases on a map are also sharable and displayable on digital maps. These maps can be zoomed for increased clarity and view of more study areas. The availability of digital maps at any part of the world has over the years contributed to the huge ambiguity of GIS in the health sector. With these vigorous functionalities, better decision making is realizable to the fullest in health-related issues. GIS can also be used to determine the relationship between Natural/ Manmade Spatial features and diseases contracted by persons living in those areas. E.g., Disease sourced from mines, water body, etc. Secure geographic information systems can be used to collect data from varying sources and perform statistical methods on the data.
- Real Estate development: Mapping in the survey, geographic statistics, and location addresses are one of the uses of GIS in real estate. A large collection of data is analyzed to help investors determine which areas would best prosper and work for real estate. It is done by even considering population and demographic features available in a given location. Land with the best characteristics can be sorted an organized as-needed together with its characteristics. Robust geographic systems may be designed to even send update information to clients who have indicated interest in real estate.
- Urban and Rural Planning: Various ideas and innovations are being brought out by GIS to ensure the construction of world-class cities, towns, and rural areas with the least congestion. Various developments like electricity, roads, railway, optic fiber, water and sewer lines which

determine how an urban or rural area are also in control of Geographic Information Systems.

- Geo-demographics: This social science involves studying a population of a given location. This population might belong to human beings or even sometimes animals. Geo-demographics specifically seek to address how many people are living at a given place over a certain time. Population maps can thus be analyzed using GIS.
- History: History is the social science that involves the study of past activities. Geographic history involves the study of formation and activities of geographic features like volcanic mountains, plateaus, hills, plains, depressions and also their interaction with Humans communities in past. All these features affected humans who existed then and must be studied to always be on the lookout today. History also involves the study and excavation of fossils and human remains. These remains are usually in most cases buried below the earth. It may thus be difficult to piece out spatially distributed useful information without the use of geographic systems like GIS. Remote sensing technology such as LIDAR, Imaging Radar is employed by Archeologist and Anthropologist to identify ancient civilization's buried colonies and cities. The Geographic Information System can be used to analyze their depths and characteristics, mapping layers with fossils records before even an actual excavation by historians who want to do further research on them. GIS is thus essential in tracking the existence of man in History.
- Emergency planning: In this field of study, much can be done through GIS. We all acknowledge that sometimes emergencies are unavoidable and wise decisions must be made to save lives. These emergencies, like flooding and fire tragedies, require a quick response and accurate mapping by relevant authorities in various governments which has been done manually. However, most governments have, with time begun to adopt GIS to help in any analyses that may be required. GIS can be integrated with weather systems to offer predictions and useful statistics that help in decision making by the government. Satellite images of places encountering disasters can alsobe sent for analysis and giving way forward to any base station on earth for processing by GIS. These systems would send help to enable navigation during evacuation operations of affected victims.
- Criminology: Criminal investigation by intelligence services has received great help from GIS. Areas with high crime rate, etc. can be mapped. Some satellites even can view very detailed real time images on earth, and this can be used to haunt those trying to escape the arm of the law. Digital images of areas displayed on computer systems of police departments

are delivered at ease for investigation operations. Deployment of police at a critical location can done by using map services Tracking Drones are being used to monitor crimes and criminals.

- Political analysis: GIS can also be used perform political opinion analysis of people living in different states, region of a country by linking opinion polls to maps, we can analyse the electoral mood of people living in different cities throughout a country.

4. Geography

- Geomorphology: Geomorphic units have been identified based on interpretation of the aerial photographs and the Survey of India's topo-sheets and the LANDSAT imageries in the various regions like Pali district, Jodhpur district, Luni basin and Tripura state.These geomorphic units have different physical potential and provide a sound base for land-use planning and mapping. Based on aerial photographs, a study of Hiran catchment (Jabalpur district), highlights the physical and hydrological characteristics of the area, Using quantitative analysis for land resource development and management. GIS has been used for soil productivity assessment and mapping. Drought monitoring is an important aspect of satellite monitoring in Maharashtra. The National Remote Sensing Agency (NRSA) carried out a survey for soil association mapping, land degradation and ground water exploration to aid drought relief in Bundelkhand region of Uttar Pradesh. A GIS service (Bhukosh) has been initiated by the Geological Survey of India for locational and distributional identification of various kinds of rocks and mineral deposits.
- Hydrology and Water Resources: Targeting ground water in hard rock, using remote sensing and GIS, is a recent phenomenon in hydrological science. In India, remote sensing is used for monitoring the aspects such as measurements of evapo- transpiration, measurement of water surface roughness, rainfall distribution and infiltration, ground water discharge and salt content of water and light absorption. Aerial photographs have been utilized to collect information regarding water storage,season and long-term fluctuations of lake and river surface aerial extent, assessment of underground and soil moisture. The weather satellite imagery is usually used to monitor ice and snow cover conditions providing important inputs for water management and flood prediction. In recent years, microwave sensor data is being used for flood monitoring during cloud cover condition. The sustainable benefits from coastal marine resources require considerable improvement in the application of remote sensing imageries to improve both utilization and management of these resources. The launching of the OCEANSAT-1 satellite has brought in significant improvement in the ocean monitoring.

- Urban Development: Increasing demands in urban planning and management sectors call for coordinate application of remote sensing and GIS for sustainable development of urban area. Availability of high-resolution data form IRS-1C and ID satellite has revolutionized the process of thematic mapping and spatial- data base creation, especially in the context of urban sprawl and regional planning. The application includes use of remote sensing in infrastructure development, urban sprawl mapping, regional planning for air and noise monitoring network, development/monitoring of roads, their management systems and landfill site selection, etc.
- Monitoring of Natural Hazards: The remote sensing data, particularly from satellites, have unique capability to monitor the condition of atmosphere and on Earth's surface. The images from an environmental satellite in conjunction with the conventional data help in monitoring severe connective situations, e.g. cyclones and storms, responsible for large-scale destruction. The visible, infra-red and enhanced infra-red images are utilized to estimate rainfall in sensitive areas. Flood plain zonation is a key tool in managing the disaster in the various parts of India. Drought monitoring is an important aspect of satellite monitoring in Drought hit states of the country. RS provides valuable data-base to study annual biomass production, agro- climatic zoning and rain-use efficiency in addition to its use in drought monitoring.

5. Biodiversity

- Data collection: Remote sensing is a useful tool in data collection. To ease the process of biodiversity conservation, remote sensors are used to take and record data of the environment, which is further used in analysis and surveys. Through this data, the environment is conserved, and natural resources are used sustainably.
- Generating maps: Remote sensors are airborne components, and therefore they usein generating maps of terrestrial ecosystems. These maps are used to show forested or non-forested areas and other physical features. The resulting map may be used in ensuring these areas converse.
- Protecting forested areas: Biodiversity can be conserved by preventing and protecting cutting down of trees in forested areas. It is tricky to monitor the thick parts of the forests from the earth's surface, and therefore remote sensing technique comes in. Since this technique is remotely used, it can capture images from the sky and thus can be used to protect the cutting of trees and helps in forest conservation.
- Monitoring natural resources sites: In order to conserve biodiversity, natural resources should be utilized efficiently. To achieve this, remote sensors are used to monitor sites that contain resources and regulate and control events of over- exploitation of resources.

- Change detention: Remote sensors have wider coverage and have the ability to detect environmental changes. Therefore, they play a major role in ecosystem and species services. Remote sensor data is used to plan for the expected change, and this facilitates timely response.
- Environmental conservation: For biodiversity to be conserved, the environment should be conserved. Remote sensors help to conserve the environment; they help in assessing the quality of air and in minimizing pollution that has a negative impact onthe environment. For example, Remote sensors measure the amount of aerosol and sprays on the air, important information that is used to curb air pollution.
- Animal conservation: Both wild and domestic animals are components of biodiversity. Thus they need to be conserved. In the case of game parks, remote sensors are useful tools, they are used to monitor all the operations taking place within the park, and this can be used to curb poaching and hence to minimize events of animal killings.
- Land-cover classification and protection: Some vegetation is valuable, however regular assessment of these plants may be difficult, thanks to remote sensing technique for easing the process. The vegetations are monitored and protected from the sky as the remote sensor are airborne components. Additionally, data from the remote sensor is used in classifying plants according to their families.

6. Natural Resource Management

- Land management: In India, the agriculture sector alone sustains the livelihood of around 50 per cent of the population (Ministry of Agriculture and Farmer's Welfare, 2016), therefore, increase in crop productivity has been the main concern. Since, the scope for increasing area under agriculture is rather limited, advanced crop production forecasting is required for better policy making. Even though globally there have been large experiments related to agriculture assessment, in India it was first attempted by the Indian Space Research Organization (ISRO) and the Indian Council of Agricultural Research (ICAR). The experiment–Agricultural Resource Inventory and Survey Experiment (ARISE) used aerial colour photographs to estimate crop acreage in Anantapur district of Andhra Pradesh and in Patiala district of Punjab. The other major uses of remote sensing and GIS in agriculture include crop identification, stress detection, identification of planting and harvest dates, crop yield modeling, estimation of pest and disease infestation, soil moisture estimation, irrigation monitoring, soil mapping, drought monitoring, land covering, land degradation mapping and more.
- Soil management: GIS is a powerful tool that is especially relevant for soil management–information which could help priorities development

actions. GIS, RS and GPS have much to offer for preparing soil fertility maps. It is widely recognized that satellite remote sensing can provide an inexpensive, rapid and effective method of data collection and production of various kinds of thematic maps. Once the soil fertility maps are created, it is possible to transform the information from Soil Test Crop Response (STCR) models into spatial fertilizer recommendation maps. Such maps provide site-specific recommendation without testing the soil. The recommendations can be obtained by an extension agent/farmer simply by locating his farm on the map. Remote sensing has evolved into an important (supplement to ground observations in the study; of terrestrial vegetation and soil.

- Watershed management: A watershed is a natural hydrological unit and its management involves the holistic linking of upstream and downstream areas. Scientists now recognize that the best way to protect natural resources is to understand and manage them on a watershed basis. Everything that is done in a watershed affects the watershed's system (Yongsheng, 2004). Watershed management provides a way for sustainability, integrating natural resource management with communities and their livelihood. They are important as ecological units from the perspective of conservation of rainfed marginal areas enabling a sustainable living. Availability of clean water is currently being regarded as a great challenge, especially in India. With the help of satellite data and GIS: water bodies such as rivers, lakes, dams and reservoirs can be mapped in 3D formats and the data can be used in the planning of sustainable management of water bodies.
- Urban land management: Urbanization is important and inevitable for development, but its proper planning and management is crucial to sustenance. One of the features of GIS is multilayered mapping, which is significant for urban planning. Urban management requires layered data on a single map wherein use of remote sensing and GIS applications comes handy. This data helps municipal corporations/ town planning boards to build cities that are better organized (Patkar, 2010; Rai and Kumra, 2011). In recent times, agencies, both central and state, have recognised the utility of GIS in problem solving through reducing cost and time. It is being actively used by municipal corporations, real estate firms, community developers and public works departments to deal with aspects like health infrastructure, disaster management, security, education, culture, transport, telecommunications and electricity etc. The information systems with socio-economic data overlaid upon satellite data makes urban planning cost effective and accurate. India used to be dependent on photogrammetry for obtaining information regarding urban planning. But since March 17, 1988 with the launch of its first satellite

(IRS-1A) equipped with LISS-I sensor, which could acquire data at a spatial resolution of 72.5 megapixels, the application of remotely sensed data (from various sensors) in urban and regional planning processes has gained a momentum. First-time national level, high-resolution satellite data utilisation for preparation of large scale (1:10,000) urban geospatial database under the National Urban Infrastructure (NUIS) programme is being done by ISRO for 152 towns (covering ~50,000 sq km) using high-resolution satellite data (Cartosat-IPAN and Resourcesat LISS-IV).

- Coastal zone management: Coastal ecosystems have high ecological significance. GIS and remote sensing data is used to study coastal ecosystem and marine living resources which include coastal habitats like mangroves, coral reefs and more. Apart from this, suspended shoreline dynamics can be studied and climatic changes leadingto cyclone and sea level rise may be of special interest too.
- Mineral resource management: Remote sensing in mineral exploration can help miners find and evaluate deposits without having to undertake massive exploration operations. Such images are used in two key operations. Such images are used in two key ways–through mapping the geology, faults and fractures of an ore deposit and by recognizing hydrothermally altered rocks by their spectral signature. Images are gathered either through optical sensors or through synthetic aperture sensors. Optical sensors measure the spectral data of sunlight reflected from the Earth's surface. Synthetic aperture sensors are able to sense electromagnetic data by transmitting microwaves and receiving the back scatter waves from the Earth's surface. Remote sensing is a valuable tool in the discovery of high-value commodities such as diamonds and gold, which are becoming more difficult to locate. While it may not show exactly where major deposits are, data gathered through sensors can be used to narrow down field surveys to smaller areas. Expensive operations like drilling and fieldwork can be undertaken after relevant satellite-based information is gathered (Kay, 2018). The greatest advancement in mineral exploration has been the ability to synthesize various forms of data. Known drill results can be integrated with topographic maps, air photos, structural maps and ore grade data. Data synthesis can greatly increase the accuracy and effectiveness of an exploration programme.

7. Urban Planning

- Resource inventory: GIS platforms, especially those used in conjunction with remote sensors, decrease time spent collecting land-use and environmental information. With remote images, urban planners can detect current land use, as well as changes to land use for an entire urban area. These images can also be used to create compelling visualizations with 3D CAD models.

- Creating land-use maps & plans: Future land-use maps act as a community's guide to future infrastructure, build plans, and public spaces. These map help ensure that a city's urban planning accounts for environmental conservation, pollution, mitigating transportation issues, and limiting urban sprawl. With GIS, urban planners can quickly create maps of the city as it is today, and then use various modeling and predictive data techniques to explore scenarios for the future. Ideally using this exercise to create a future land-use map that is thoughtful, sustainable, and sound.
- Planning applications: GIS can help the government and businesses process and organize planning applications. Many GIS portals can be made public facing, which means citizens can access data such as parcel outlines and information, county/district boundaries, and area zoning. With vital information more widely available to all, government resources (which might have been spent fielding these requests and finding the data) can be put to use elsewhere. Moreover, with all the applications stored in a central database, organization, processing, and status tracking becomes much simpler.
- Analyzing socioeconomic & environmental data: Creating future land-use maps must take into account several environmental scenarios, as well as project future demand for land resources. Modeling must include population data, economic activities, and spatial distribution. The visual component of GIS makes analyzing location-based data (like socioeconomic and environmental trends) simpler and moreeffective. GIS enables the creation of thematic maps i.e., maps that combine data and location in order to explore correlation and display trends. With the various data sets stored in the GIS database, users can create layered images that include topography, street maps, thematic maps, and more - helping to easily identify ideal spaces, as well as areas of potential conflict.

8. Disaster Management

The application of remote sensing and GIS has become a well-developed and successful toolin disaster management, as our location observation programmes and the requisite for hazard mitigation and monitoring rank high in the planning of new satellites. GIS allows for the combination of different kinds of data using models. It allows for the combination of the different kinds of spatial data with non-spatial data, attribute data and use them as useful information in the various stages of disaster management. Various disasters like earthquake, landslides, flood, fires, tsunamis, volcanic eruptions, and cyclones are natural hazards that kill lots of of people and destroy property and infrastructure every year. The rapid increase of the population and its increased concentration, often in hazardous environment, has escalated both the frequency and severity of natural disasters. Among the tropical climate and unstable land forms, coupled with deforestation, unplanned growth propagation non- engineered constructions which make the disaster-prone areas sheer vulnerable, slow communication, poor budgetary allocation for disaster prevention, developing countries suffer more or less unceasingly by natural disasters.

- Disaster mapping: Disaster mapping is the drawing of areas disturbed through excessive natural or manmade troubles resulting in loss of life, property and national infrastructures. It is normally possible to define the area affected by the disruption. The delineation can occur through the use of ground-based observations or through the use of remote sensing devices such as aerial photographs or satellite images. From the information gathered, it is possible to map the affected areas and provide information to the relief supplying groups. Disaster mapping is a tool for assessing, storing and conveying information on the geographical location and spread of the effects, or probable effects of disasters. The difficulty with traditional manual maps is that they are tedious and time consuming to prepare, difficult to update and inconvenient to maintain. Remote sensing is emerging as a popular means of map preparation while GIS can be used for storage, analysis and retrieval. Under remote sensing techniques, maps can be prepared using satellite data or aerial photographs and then digitised and stored on computers using GIS software. Disaster maps generally show risk zones as well as disaster impact zones. These are marked areas that would be affected increasingly with the increase in the magnitude of the disaster. These could include landslide hazard maps,flood zone maps, seismic zone maps, forest fire riskmaps, industrial risk zone maps etc.

Important GIS Packages

Some of the important GIS packages are;

1. *ARC/INFO* is one of the first GIS packages that was available commercially and is a package used all over the world. It has been developed by Environmental System Research Institute, USA. It is available on wide range of platforms-PC's, workstations, PRIME systems. It is also available on variety of operating systems- DOS, UNIX, VMS etc.
2. *PAMAP* is a product of Graphic Limited, Canada and is an integrated group of software products designed for an open system environment. The package is modular and is designed to address the wide range of mapping and analysis requirements of the natural resource sector. It is also available on variety of platforms - Pentium, 486/386/286 PC's, UNIX, SUN, VAX system.
3. *MAPINFO* is a popular package translated in to several languages and ported to several platforms like Windows, Macintosh, Sun, and HP workstations.
4. *GRASS,* a public domain UNIX package with large established user base which actually contributes to the code that is incorporated in to new versions.

5. *ISROGIS* is a state-of-art GIS package with efficient tools of integration and manipulation of spatial and non-spatial data and consists of a set of powerful modules. It is available on PC platforms on MS-Windows and on UNIX and SUN platforms.
6. *IDRISI* has been developed by Clarke University, USA, and an inexpensive PC based advanced features including good import export facility, a new digitization module, and some image processing facilities.
7. *GRAM* is a PC based GIS tool developed by IIT, Bombay. It can handle both vector and raster data and has functionality for raster-based analysis, image analysis etc.
8. *QGIS:* A Free and Open-Source Geographic Information System developed by Gary Sherman in early 2002. QGIS is maintained by volunteer developers who regularly release updates and bug fixes. As of 2012, developers have translated QGIS into 48 languages and the application is used internationally in academicand professional environments. Several companies offer support and feature development services.

References

1. Anand A, Kumari, B., Nayak S.R., Murthy, and Krishna,Y.V.N., 2005.Locating oceanic Tuna resources in the eastern Arabian sea using remote sensing.Journal of Indian Society of remote sensing, 33(4), pp.,512-520
2. Lutchman, I., 2003. New technical approaches in fisheries management: the precautionary approach and the ecosystem approach. In De Fontaubert, C. & I. Lutchman (eds), Achieving Sustainable Fisheries: Implementing the New International Legal Regime. IUCN, The World Conservation Union: 31–45.
3. Madhavan N., Thirumalai V. D., Ajith J. K. and Sravani K., 2015. Prediction of Mackerel Landings Using MODIS Chlorophyll- A, Pathfinder SST, and SeaWiFS PAR. Indian Journal Of Natural Sciences,5(29), pp.4858-4871
4. Pierce, G. J., J. Wang & Valavanis, V. D., 2002. Application of GIS to cephalopod fisheries: workshop report. Bulletin of Marine Science 71: 35–46
5. Valavanis , V.D., Pierce, G.J., Zuur,A.F., Palialexis,A., Saveliev,A., Katara,I. and Wang, J.,2008.Modelling of essential fish habitat based on remote sensing, spatial analysis and GIS. Hydrobiologia, 612:5–20
6. Valavanis, V. D., Georgakarakos, S., Kapantagakis, Palialexis, A., & Katara, I., 2004. A GIS environmental modelling approach to Essential Fish Habitat Designation. Ecological Modelling, 178: 417–427.
7. APDRc website (httpJ/apdrc.soest.hawaii.edui)
8. NASA-GioVANNI (httpsi/giovanni.gsfc.nasaugov/glovanniDl)
9. NASA- Ocean Colour (https.J/oceancolor.gsfc.nasa.govJ)

CHAPTER 2

SPATIAL ANALYSIS AND IMAGE PROCESSING

Vinod Kumar Yadav, Palsam Kartik K Goud, Mahesh Sharma and Liton Paul

Introduction

Spatial analysis is the process of extracting or creating new information about a set of geographic features to perform routine examination, assessment, evaluation, analysis or modelling of data in a geographic area based on pre-established and computerized criteriaand standards. It is the process that provides general ability to manipulate spatial data into different forms and extract additional meaning. The first attempts of spatial data analysis date back to the 1960s and belong to Canada and the earliest objective for GIS applications was the systematization of the country's natural resources and ever since spatial analysis in GIS has expanded worldwide. Spatial analysis allows determining if the selected area is suitable for particular business activities, detecting changes, identifying tendencies, assessingrisks, outlining possible outcomes, and preventing losses. This type of analysis has proven to be highly effective for evaluating the geographic suitability of certain locations for specific purposes, estimating and predicting outcomes, interpreting and understanding change, detecting important patterns hidden in your information, and much more. There are many ways of defining spatial analysis, but all in one way or another express the fundamental idea that information on locations is essential. It is a diverse and comprehensive capability that includes the simple visual analysis of maps and imagery, computational analysis of geographic patterns, finding optimum routes, site selection, and advanced predictive modelling. Basically, think of spatial analysis as "a set of methods whose results change when the locations of the objects being analyzed change". Examples of spatial analysis include measuring distances and shapes, setting routes and tracking transportations, establishing correlations between objects, events, and places via referring their locations to geographical positions (both live and historical). For instance, calculating the average income for a group of people is not spatial analysis because the result doesn't depend on the locations of the people whereas calculating the centre of the India's population, however, is spatial analysis because the result depends directly on the locations of residents.

Major categories and utilities in spatial analysis

1. **Understanding where:** This includes geocoding your data, putting it on a map, and symbolizing it in ways that can help you visualize and understand your data. This step includes:
    - Understanding where things are (location maps): One of the simplest questions wecan answer with spatial analysis is, where are things located? This could be a simple store directory map at the mall or a more sophisticated map showing real-time positions of moving vehicles and events
    - Understanding where the variations and patterns in values are (comparative maps): Comparative maps take advantage of different approaches to symbology (colour, size, dot density, charts) and allow you to analyze and further understand spatial variations.With comparative mapping you can visualize the highs and lows and their distribution across space. These comparative analyses can be accomplished using historical, current, and even real-time analytical maps.
    - Understanding where and when things change: The world is constantly changing. Mapping the changing conditions in a place over time, such as loss of vegetation, can help us anticipate future conditions and implement policies that will positively impact our world.
2. **Measuring size, shape, and distribution:** To know how large an object is, or you may want to describe an object in terms of its geometric properties, such as area, perimeter, length, height, and volume. It involves:
    - Calculating individual feature geometries: Spatial analysis is not just about understanding where things (features) are located, it is also about the spatial characteristics of these features. For example: How big is the lake? How long is the trail?
    - Calculating building heights from satellite imagery, calculating the volume of a lake, and calculating the slope of a bike path are all examples of spatial analysis.
    - Calculating geometries and distributions of feature collections:

Although individual features have their own geometric properties (size and shape), a collection of features takes on additional properties such as extent and distribution

3. **Determining how places are related:** It includes a set of questions that help describe and quantify the relationships between two or more features. This involves
   - Determining what is nearby or coincident: It is understanding the spatial relationship of distance or proximity is a common spatial analysis process. Distances can be measured, and features can be selected based on their distance characteristics.
   - Determining and summarizing what is within an area(s): Spatial aggregation provides a mechanism to count and summarize data over geographic areas.
   - Determining what is closest: It is done by using different spatial analysis approaches to determine what is closest in space or time.
   - Determining what is visible from a given location(s): The important relationship between places is intervisibility; the ability to calculate sight lines and viewsheds, taking into account of the terrain, the built environment, and the earth's curvature.
   - Determining overlapping relationships in space and time: Two or more features can have a variety of relationships in space. Are they close, nearby, visible, accessible, and so on? However, the world is not static—locations, properties, and relationships may change over time. Bringing time into the analysis adds yet another dimension to all the relationships.
4. **Finding the best locations and paths:** A very common type of spatial analysis, and probably the one you are most familiar with, is optimization and finding the best of something. It can be the best route to travel, the best path to ride a bicycle, the best corridor to build a pipeline, or the best location to site a new store. This involves:
   - Finding the best locations that satisfy a set of criteria
   - Finding the best allocation of resources to geographic areas
   - Finding the best route, path, or flow along a network
   - Finding the best route, path, or corridor across open terrain
   - Finding the best supply locations given known demand and a travel network
5. **Detecting and quantifying patterns:** This analysis includes mathematically detecting andquantifying patterns in data. For example, spatial statistics can be used to find hot spots and outliers; data mining techniques can

be used to find natural data clusters; and both approaches can be used to analyze changes in patterns over time. This involves:

- Where are the significant hot spots, anomalies, and outliers?
- What are the local, regional, and global spatial trends?
- Which features/pixels are similar, and how can they be grouped together?
- Are spatial patterns changing over time?

6. **Making predictions:** These techniques can be used to predict and interpolate data values between sample points, find the factors related to complex phenomena, and make predictions in the future or over new geographies. It involves:
   - Given a success case, identifying, ranking, and predicting similar locations: Given a set of successful locations, we can find other locations, with similar characteristics, that are likely to succeed; and we can assess which locations are more likely to fail.
   - Finding the factors that explain observed spatial patterns and making predictions: As we try to understand the factors related to an observed pattern, it is often necessary to explore a number of variables, test their significance, and then analyze various combinations so that we can use these models to predict the same phenomena in otherareas or other times using spatial regression
   - Interpolating a continuous surface and trends from discrete sample observations: What do you do when you have sample data that does not cover the entire area of interest? Predictive spatial analysis can be used to predict, or interpolate, between discrete data points to create a continuous surface.
   - Predicting how and where objects spatially interact (attraction and decay): These predictive models use the concept of mass and gravity to predict the behaviour of supply and demand for retail site selection and for other application areas.
   - Predicting where phenomena will move, flow, or spread: Many geographic phenomena have the ability to move, flow, or spread—a forest fire, a disease outbreak, a chemical contamination, or a marketing campaign. Predicting how these phenomena move in space and time is an important type of spatial analysis that involves modelling complex systems, interactions, behaviours, and responses.
   - Predicting what-if: Models are used to predict outcomes, help us understand our world, and help people make informed decisions. One of the most valuable aspects of modelling is the ability to test

and evaluate potential actions and their resulting impacts. Using a what-if, iterative approach, many different scenarios can be evaluated before taking action.

Types of spatial analysis

Types of spatial analysis vary from simple to sophisticated. Mainly, spatial analysis can be divided into six categories: queries and reasoning, measurements, transformations, descriptive summaries, optimization, and hypothesis testing.

1. **Queries and reasoning** are the most basic of analysis operations, in which the GIS is usedto answer simple questions posed by the user. No changes occur in the database and no new data are produced.
2. **Measurements** are simple numerical values that describe aspects of geographic data. They include measurement of simple properties of objects, such as length, area, or shape, and of the relationships between pairs of objects, such as distance or direction.
3. **Transformations** are simple methods of spatial analysis that change data sets by combining them or comparing them to obtain new data sets and eventually new insights. Transformations use simple geometric, arithmetic, or logical rules, and they include operations that convert raster data to vector data or vice versa. They may also create fields from collections of objects or detect collections of objects in fields.
4. **Descriptive summaries** attempt to capture the essence of a data set in one or two numbers. They are the spatial equivalent of the descriptive statistics commonly used in statistical analysis, including the mean and standard deviation.
5. **Optimization techniques** are normative in nature, designed to select ideal locations for objects given certain well-defined criteria. They are widely used in market research, in the package delivery industry, and in a host of other applications.
6. **Hypothesis testing** focuses on the process of reasoning from the results of a limited sample to make generalizations about an entire population. It allows us, for example, to determine whether a pattern of points could have arisen by chance based on the information from a sample. Hypothesis testing is the basis of inferential statistics and forms the core of statistical analysis, but its use with spatial data can be problematic.

Levels of Measurement in Spatial Analysis

1. **Nominal:** Categories, states, or names of things.

   Examples: ID numbers, eye colour, hair colour, marital status, occupation

2. **Ordinal:** Values have a meaningful order (ranking) but magnitude between successive values are not known. Examples: rankings (e.g., taste of potato chips on a scale from 1-10),grades, height in {tall, medium, short}, size {small, medium, large}
3. **Interval:** All variables that are measured in an interval scale can be added, subtracted, multiplied but calculating ratio is not possible. Examples: calendar dates, temperatures in Celsius or kelvin
4. **Ratio:** Ratio scale has all the characteristics of an interval scale, in addition, to be ableto calculate ratios. Examples: temperature, length, time, and count

Uncertainties and Errors in Spatial analysis

- **Spatial uncertainty:** When objects do not have a discrete, well-defined extent. They may have indistinct boundaries, impacts that extend beyond their boundaries or may simply be statistical entities. The attributes ascribed to spatial objects may also be subjective.
- **Vagueness:** The criteria that define an object as x are not explicit or rigorous.
- **Ambiguity:** occurs when y is used as a substitute, or indicator, for x because x is not available.
- **Physical measurement error:** Instruments and procedures used to make physical measurements are not perfectly accurate.
- **Digitizing error:** Digitizing, or the electronic tracing of paper maps, is prone to human error. Lines may be drawn too far, not far enough, or missed entirely. Errors caused by digitizing mistakes can be partially, but not completely, fixed by software.
- **Ecological fallacy:** The ecological fallacy is the mistake of assuming that an overall characteristic of a zone is also a characteristic of any location or individual within the zone.

Interpolation

Environmental and Climatic data are used for scientific and technical purposes in many different areas of environmental research and resource management, hydrology and agriculture. Climate series are often limited in spatial and temporal coverage. The management and maintenance of gauge stations to obtain reliable data can be very expensiveand problematic. In order to tackle these problems, it is useful to develop spatial interpolation techniques to assess climatic variables at ungauged sites. In general, spatial interpolation techniques use observations of the same variable made at other sites.

Inverse Distance Weighting (IDW)

Inverse Distance Weighting (IDW) is an interpolation technique in which interpolated estimates are made based on values at nearby locations weighted only by distance from the interpolation location. IDW does not make assumptions about spatial relationships except thebasic assumption that nearby points ought to be more closely related than distant points to the value at the interpolate location. This technique determines cell values using a linearly weighted combination of a set of sample points. The weight is a function of inverse distance.IDW allows the user to control the significance of known points upon the interpolated values, based upon their distance from the output point.

Kriging

Kriging provides a means of interpolating values for points not physically sampled using knowledge about the underlying spatial relationships in a data set to do so. Variograms provide this knowledge. Kriging is based on regionalized variable theory which provides an optimal interpolation estimate for a given coordinate location, as well as a variance estimate for the interpolation value. It involves an interactive investigation of the spatial behavior of the phenomenon before generating the output surface. It is based on the regionalized variable theory, which assumes that the spatial variation in the phenomenon is statistically homogeneous throughout the surface; that is, the same pattern of variation can be observed atall locations on the surface. This hypothesis of spatial homogeneity is fundamental to the regionalized variable theory. Data sets known to have spikes or abrupt changes are not appropriate for the Kriging technique. In some cases, the data can be pre-stratified into regions of uniform surface behaviour for separate analysis. Kriging is based on the regionalized variable theory that assumes that the spatial variation in the phenomenon represented by the z-values is statistically homogeneous throughout the surface (for example,the same pattern of variation can be observed at all locations on the surface). This hypothesisof spatial homogeneity is fundamental to the regionalized variable theory. Point sets that are known to have anomalous pits or spikes, or abrupt changes that might be represented by break lines in a TIN, are not appropriate for the kriging technique. In some cases, the datacan be pre-stratified into regions of uniform surface behaviour for separate analysis. The spatial variation is quantified by the semi variogram. The semi variogram is estimated by the sample semi variogram, which is computed from the input point dataset. The value of the sample semi variogram for a separation distance of h (referred to as the lag) is the average squared difference in z-value between pairs of input sample points separated by h. The sample semi variogram is calculated from the sample data with the equation:

$$\gamma(h) = \frac{1}{2n}\sum_{i=1}^{n} \{Z(xi) - Z(xi + h)\}^2$$

where n is the number of pairs of sample points separated by the distance h.Figure 2.1is an example of a typical semi variogram:

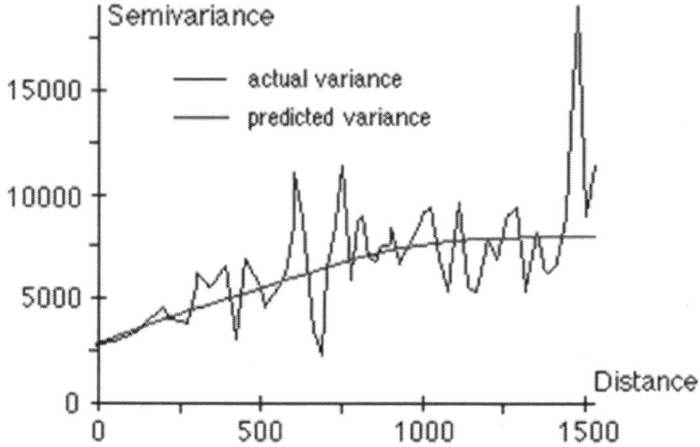

Figure 2.1: Semi variogram

There are several important features worth noting in the plot of the sample semi variogram. At relatively short lag distances of h, the semi variance is small but increases with the distance between the pairs of sample points. At a distance referred to as the range, the semi variance levels off to a relatively constant value referred to as the sill. This implies that beyond this range distance, the variation in z-values is no longer spatially correlated. Within the range, the z-value variation is smaller when the pairs of sample points are closer together. The extent of the x-axis of the semi variogram is determined by the distance between the most widely separated pair of points in the input sample data. The maximum distance separating the pairs of points used to fit the mathematical function is determined by {radius}. Kriging offers two types of surface estimators: Ordinary Kriging and Universal Kriging. Ordinary Kriging is represented by the Spherical, Circular, Exponential, Gaussian, and Linear types. With these options, Kriging uses the mathematical function specified with the semi Variogram Type argument to fit a line or curve to the semi variance data in the semi variogram. Ordinary Kriging assumes that the variation in z-values is free of any structural component (drift). These five models are provided to ensure that the necessary conditions of the variogram model are satisfied.

The variance is calculated based on the average variance of all point pairs within each interval of the cell size. The variogram is then fitted to the variance points using the Levenberg–Marquardt Method of nonlinear least squares approximation. A minimum of three points (a variance value at three distances) is required for the fit.

By increasing the cell size, you will increase the number of sample points per cell size interval, thereby providing enough data points to estimate the semi variogram. Once the semivariogram is estimated, a smaller cell size can be used in creating the actual output raster.

In some cases, you may suspect that the spatial variation in z-values in the data contains local trends. Some researchers have discovered that the presence of locally changing linear drift in the data is indicated when the semi variogram has a gently parabolic concave-upward shape near the origin. Universal Kriging, represented by the Universal1 and Universal2 methods, assumes that the spatial variation across the surface also has a structural componentreferred to as drift. Drift is a systematic change in the z-values in a particular direction.

Universal Kriging assumes that the spatial variation in z-values is the sum of three components: a structural component (drift), a random but spatially correlated component, and random noise representing the residual error. The structural component represents a constant trend over the surface. The random noise is assumed to be spatially independent andhave a normal distribution. Once the structural effects have been accounted for, the remaining variation is spatially homogeneous such that the z-value difference between input sample points is merely a function of the distance between them as with Ordinary Kriging.

The Universal option uses a linear function to model the drift. The Universal2 option uses a quadratic equation to model the drift. Both methods use the linear technique to model the remaining semi variance. The Radius argument is used to select the number of sample pairs analysed for each output raster cell. The number of sample points must be large enough to allow the drift to be properly calculated. When specifying the radius, choose a distance that is within the well-correlated portions of the range.

Mathematical models

Below Figure 2.2 to 2.6 are the general shapes and the equations of the mathematical models used to describe the semi variance.

SPHERICAL

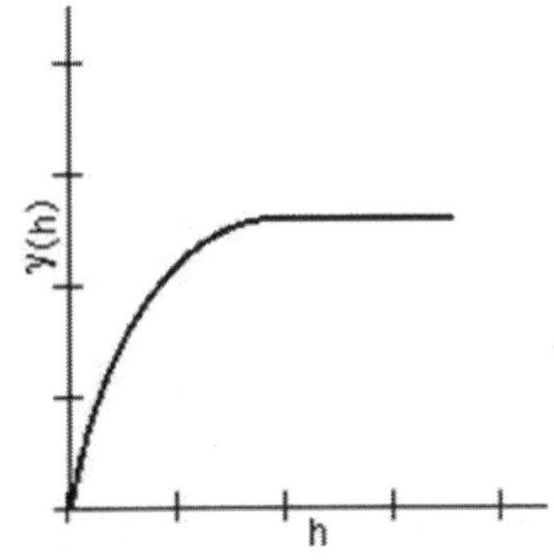

$$\gamma(h) = c_0 + c\left(\frac{3h}{2a} - \frac{1}{2}\left(\frac{h}{a}\right)^3\right) \quad 0 < h \le a$$

$$\gamma(h) = c_0 + c \quad h > a$$

$$\gamma(0) = 0$$

Figure 2.2: Shape and the equations of the spherical mathematical models

CIRCULAR

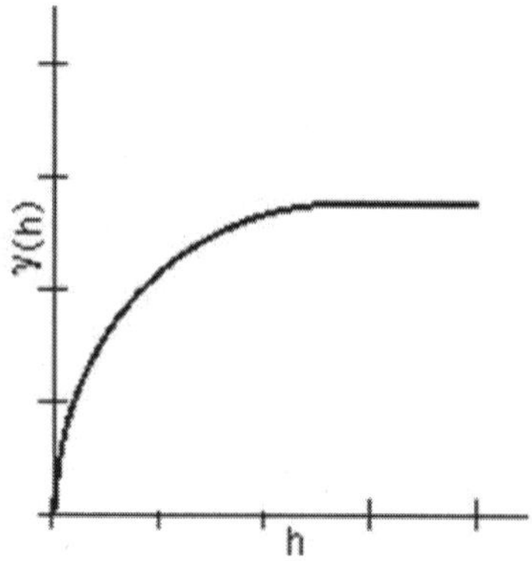

$$\gamma(\mathbf{h}) = c_0 + c\left(1 - \frac{2}{\pi}\cos^{-1}\left(\frac{h}{a}\right) + \sqrt{1 - \frac{h^2}{a^2}}\right)$$

$$0 < h \leq a$$

$$\gamma(\mathbf{h}) = c_0 + c \qquad h > a$$

$$\gamma(0) = 0$$

Figure 2.3: Shape and the equations of the circular mathematical models

EXPONENTIAL

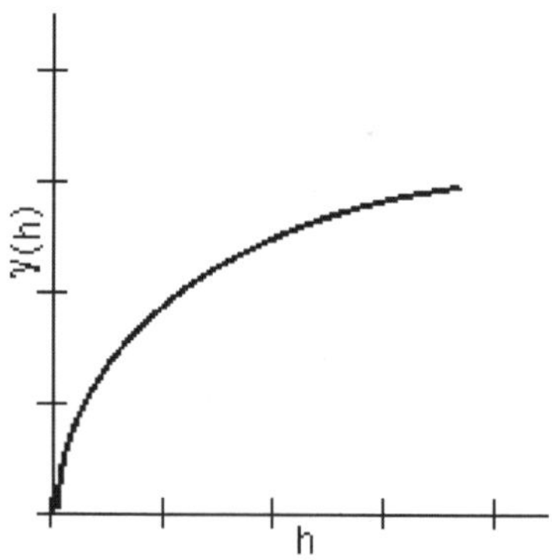

$$\gamma(\mathbf{h}) = c_0 + c\left(1 - exp\left(\frac{-h}{r}\right)\right) \qquad h > 0$$

$$\gamma(0) = 0$$

Figure 2.4: Shape and the equations of the exponential mathematical models

GAUSSIAN

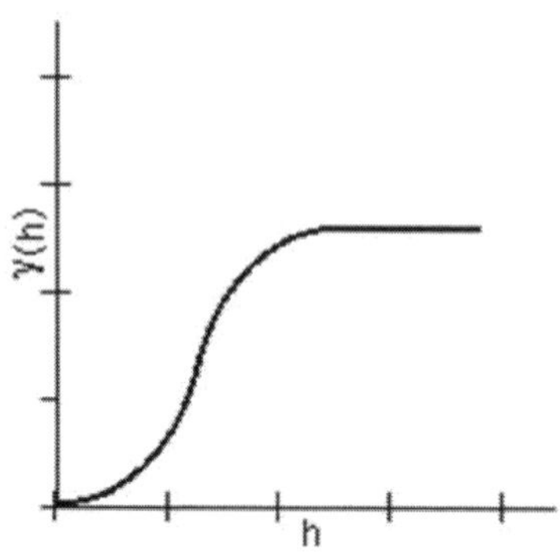

$$\gamma(\mathbf{h}) = c_0 + c\left(1 - exp\left(-\frac{h^2}{r^2}\right)\right) \qquad h > 0$$

$$\gamma(0) = 0$$

Figure 2.5: Shape and the equations of the gaussian mathematical models

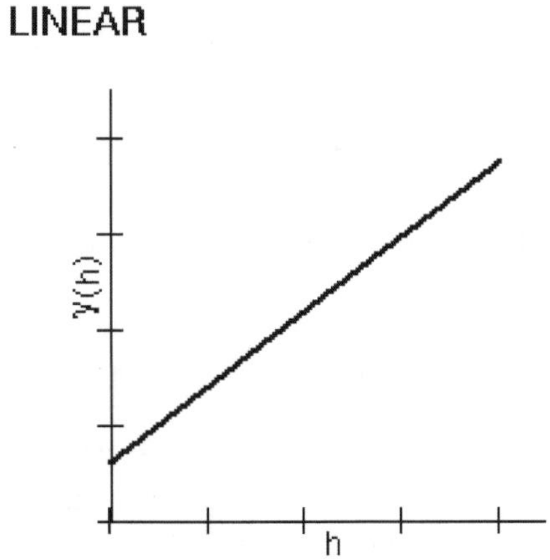

$$\gamma(\mathbf{h}) = c_0 + c\left(\frac{h}{\alpha}\right) \qquad 0 < h \leq \alpha$$

$$\gamma(\mathbf{h}) = c_0 + c \qquad h > \alpha$$

$$\gamma(0) = 0$$

Figure 2.6: Shape and the equations of the linear mathematical models

Overlay method

Overlay methods superimposes multiple data sets (representing different themes) together forthe purpose of identifying relationships between them. An overlay creates a composite map by combining the geometry and attributes of the input data sets. Tools are available in most GIS software for overlaying both Vector or raster data.

Weighted Overlay

The Weighted Overlay tool is implemented within a single tool in several steps in the general overlay analysis process. The tool consists of several stages which combines the following steps:

1. Reclassifies values in the input raster into a common evaluation scale of suitability orpreference or some similarly unifying scale
2. Multiplies the cell values of each input raster by the raster's weight of importance
3. Adds the resulting cell values together to produce the output raster.

The higher values result of overlay generally indicates that a location is more suitable. The weighted overlay tool scales the input data on a defined scale and weights of theinput raster's, added together. Weighted overlay assumes that more favourable factors resultin the higher values in the output raster, therefore identifying these locations as being the best is considered.

Image Processing

Introduction

In today's advanced technological environment, where the majority of remote sensing data is stored in digital format, practically all image interpretation and analysis involve some type of digital processing. Digital image processing can include a variety of techniques such as data formatting and correction, digital enhancement to aid visual interpretation, and even computer-assisted classification of targets and characteristics. Data must be recorded and available in a digital form appropriate for storage on a computer tape or disc in order to process remote sensing imagery digitally. Obviously, a computer system, also known as an image analysis system, with the necessary hardware and software to analyse the data is also required for digital image processing. Several commercially available software systems for remote sensing picture processing and analysis have been created. For discussion purposes, most of the common image processing functions available in image analysis systems can be categorized into the following four categories:

- Preprocessing
- Image Enhancement
- Image Transformation
- Image Classification and Analysis

Preprocessing: Preprocessing, which are commonly categorized as radiometric or geometric corrections, are normally necessary prior to the major data processing and extraction of information. Correcting the data for sensor abnormalities and undesirable sensor or ambient noise, as well as transforming the data to appropriately represent the reflected or emitted radiation detected by the sensor, are all part of radiometric corrections. Geometric correctionsinclude correcting for geometric distortions caused by sensor-Earth geometry changes and converting the data to real-world coordinates on the Earth's surface (latitude and longitude).

Georeferencing is a necessary process that all remotely sensed images must undergo, if the images are to be used in conjunction with other data sources. Only by transforming the raw images into a common coordinate system are we able to combine the imagery from one sensor with images collected by another sensor, and to combine remotely sensed images withall of the other spatial datasets being produced (such as digital elevation models, transportation maps, etc.).

Image Enhancement: Simple picture enhancements, such as 'destriping' an image, are frequently required since many sensors develop 'glitches' over time that, while not rendering the sensor incapable of providing any data, result in images with data fluctuations. That is, there may be variations in radiometric values that are unrelated to what the sensor is perceiving (i.e., unconnected to the target), but are instead the product of a sensor malfunction. Any classification we conduct on the image would combine these (meaningless) internally-derived variations in the digital numbers with the (meaningful) externally-driven changes in the spectral response patterns that we are interested in until we remove these 'internal' variations. The objective of the image enhancement is solely to improve the appearance of the imagery to assist in visual interpretation and analysis. Examples of enhancement functions include contrast stretching to increase the tonal distinction between various features in a scene, and spatial filtering to enhance (or suppress) specific spatial patterns in an image.

Image Transformation: Image transformations are procedures that are conceptually comparable to image enhancing techniques. Image transformations, on the other hand, are frequently combined processing of data from many spectral bands, unlike image enhancement techniques, which are usually done to a single channel of data at a time. To combine and alter the original bands into "new" images that better display or highlight certain characteristics in the scene, arithmetic operations (i.e. subtraction, addition, multiplication, division) are performed. We'll look at some of these procedures, such as various spectrum or band ratioing approaches and a procedure called principal components analysis, which is used to better represent information in multichannel imaging.

Image Classification and Analysis: Image classification and analysis operations are used to digitally identify and classify pixels in the data. Classification is usually performed on multi-channel data sets (A) and this process assigns each pixel in an image to a particular class or theme (B) based on statistical characteristics of the pixel brightness values. There are a variety of approaches taken to perform digital classification. We will briefly describe the twogeneric approaches which are used most often, namely supervised and unsupervised classification. Image classification is a computer vision process that can classify an image based on its visual information. For example, an image classification algorithm may be created to determine whether or not an image contains a human figure. While humans can easily discern objects, computer vision applications still face challenges in image classification.

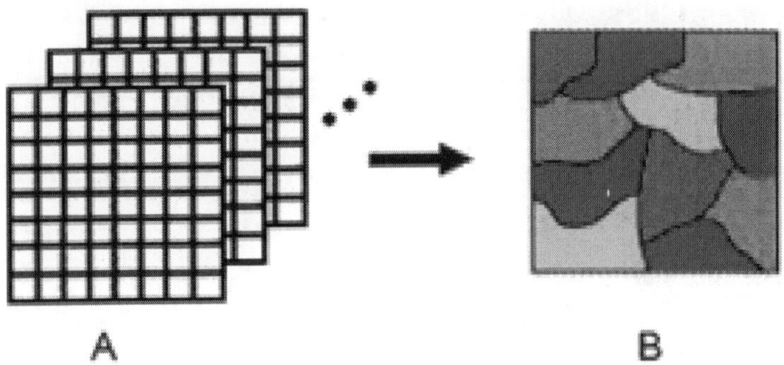

Figure 2.7: Image classification

The technique of assigning land cover classifications to pixels is known as image classification. Water, urban, woodland, farm, and grassland are examples of classes.

The process of allocating pixels in a picture to categories or classes of interest is known as image classification. Built-up places, water bodies, green foliage, barren soil, rocky areas, cloud, and shadow are some examples. The relationship between the data and the classes intowhich they are classified must be properly understood in order to classify a piece of data intodifferent classes or categories. To accomplish this with a computer, it must first be (i) trained. (ii) Classification approaches were originally created (iii) out of research in the Pattern Recognition field.

Image classification is a method of classifying all pixels in a topographical image into land cover classes automatically. Normally, multispectral data are used to classify the spectral pattern present in the data, with each pixel serving as the numerical basis for categorization. Pattern Recognition is the broad subject that deals with this concept. Spectral pattern recognition refers to a group of classification algorithms that use pixel-by-pixel spectral data to classify land cover automatically. Spatial pattern recognition involves the categorization of image pixels on the basis of the spatial relationship with pixels surrounding them. Image classification techniques are grouped into two types, namely supervised and unsupervised. The classification process may also include features, such as, land surface elevation and the soil type that are not derived from the image.

Steps in image classification:

Step 1: Definition of Classification Classes Depending on the objective and the characteristics of the image data, the classification classes should be clearly defined.

Step 2: Selection of Features to discriminate between the classes should be established using multispectral or multi-temporal characteristics, colour, textures etc.

Step 3: Sampling of Training Data Training data should be sampled in order to determine appropriate decision rules. Classification techniques such as supervised or unsupervised learning will then be selected on the basis of the training data sets.

Step 4: Finding of proper decision rule Various classification techniques will be compared with the training data, so that an appropriate decision rule is selected for subsequent classification.

Step 5: Classification depending upon the decision rule, all pixels are classified in a single class. There are two methods of pixel-by-pixel classification and per-field classification, withrespect to segmented areas.

Step 6: Verification of Results. The classified results should be checked and verified for theiraccuracy and reliability.

The 2 main image classification techniques in remote sensing are:

- Unsupervised image classification
- Supervised image classification

Unsupervised and supervised image classification is the two most common approaches. However, object-based classification has gained more popularity because it's useful for high-resolution data. To classify an image, supervised classification uses the spectral signatures derived from training examples. You may quickly construct training samples to represent the classes you wish to extract using the Image Classification toolbar. The training samples can also be used to produce a signature file, which is subsequently used by multivariate classification techniques to classify the image.

Supervised classification

Supervised classification is based on the premise that a user can select sample pixels in an image that represent specific classes, and then instruct image processing software to utilize these training sites as references for the categorization of all other pixels in the picture. The user's knowledge is used to determine training sites (also known as testing sets or input classes). The user can additionally specify how similar other pixels must be in order to groupthem together. The spectral properties of the training area, plus or minus a particular increment (typically based on "brightness" or strength of reflection in specific spectral bands), are frequently used to determine these limitations. The number of classifications into which the image is categorized is likewise determined by the user. To create final output analysis and classified maps, many analysts combine supervised and unsupervised classification procedures. The user or image analyst "supervises" the pixel categorization process in supervised classification. The user assigns the various pixel values or spectral signatures to each class. This is accomplished by identifying representative sample sites of a known cover type known as Training Sites or Areas. The spectral signatures from these training areas are then used by the computer program to classify the entire image. The classes should ideally not overlap or only overlap minimally with other classes.

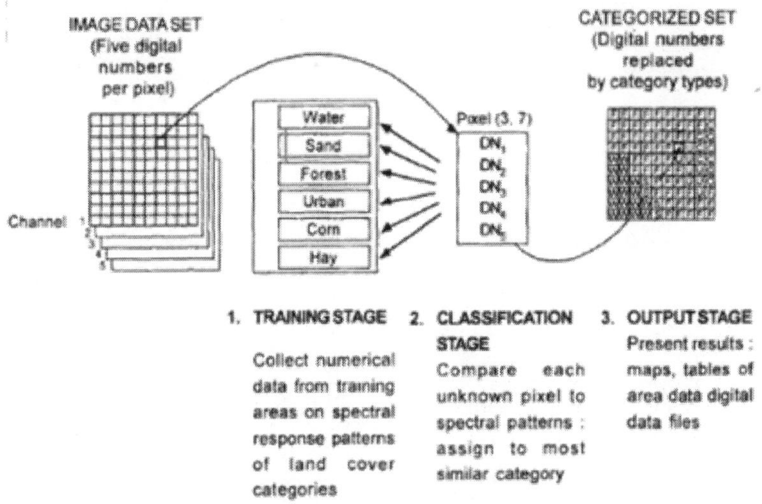

Figure 2.8: Image classification process

The following classification strategies use these "training set" descriptions of the category spectral response patterns as interpretation keys by which pixels of unidentified cover type are categorized into their classes:

Minimum-Distance Classifier: The minimum distance classifier is used to classify unknown image data to classes which minimize the distance between the image data and the class in multi-feature space. The distance is defined as an index of similarity so that the minimum distance is identical to the maximum similarity.

Advantages: It is mathematical simple and computationally efficient.

Limitations: It is insensitive to different degrees of variance in the spectral response data. Not suitable for application where spectral classes are close to one another in the measurement space and have high variance.

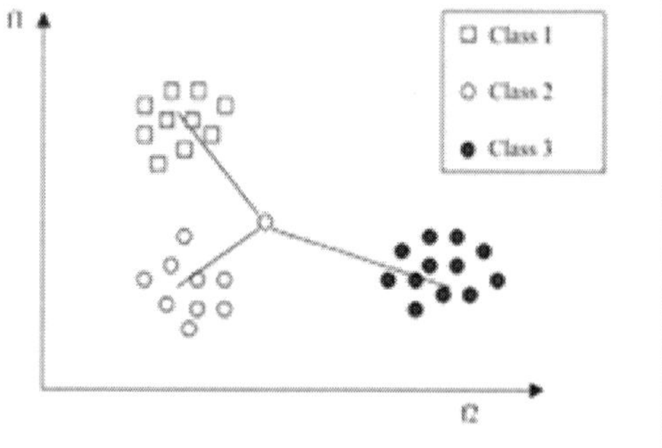

Figure 2.9: Working of minimum distance classifier

Parallelepiped Classifier: Parallelepiped classification uses a simple decision rule to classify multispectral data. The decision boundaries form an n-dimensional parallelepiped classification in the image data space. The dimensions of the parallelepiped classification are defined based upon a standard deviation threshold from the mean of each selected class. If a pixel value lies above the low threshold and below the high threshold for all n bands being classified, it is assigned to that class. If the pixel value falls in multiple classes, ENVI assigns the pixel to the first class matched. Areas that do not fall

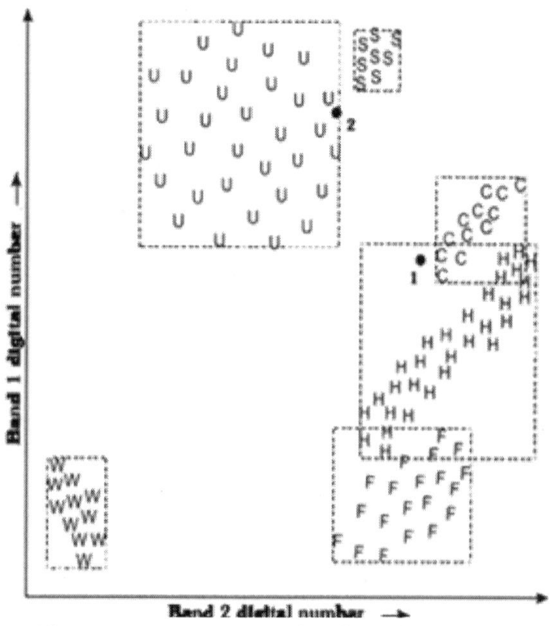

Figure 2.10: Decision rule of parallelepiped classifier for classification

within any of the parallelepiped classes are designated as unclassified. In example, the —corn□ and —hay□ categories have positive covariance and water exhibits negative covariance.

The urban class shows a lack of covariance, resulting in a nearly circular distribution. In the presence of covariance, the rectangular decision regions fit the category training data very poorly, resulting in confusion for a parallelepiped classifier. For example, the insensitivity to covariance would cause pixel 1 to be classified as "hay" instead of "corn."

Gaussian Maximum Likelihood Classifier: The maximum likelihood classifierquantitatively evaluates both the variance and covariance of the category spectral response patterns when classifying an unknown pixel. It is assumed that the distribution of the cloudof points forming the category training data is Gaussian (normally distributed). Here, thedistribution of a category response pattern can be completely described by the mean vector and the covariance matrix. The probability density functions are used to classify anunidentified pixel by computing the probability of the pixel value belonging to eachcategory. After evaluating the probability in each category, the pixel is assigned to the most likely class (highest probability value) or can be labelled as unknown' if the probabilityvalues are all below a threshold set by the analyst.

Unsupervised classification

Unsupervised classification is where the outcomes (groupings of pixels with common characteristics) are based on the software analysis of an image without the user providing sample classes. The computer uses techniques to determine which pixels are related and groups them into classes. The user can specify which algorism the software will use and the desired number of output classes but otherwise, does not aid in the classification process. However, the user must have knowledge of the area being classified when the groupings of pixels with common characteristics produced by the computer have to be related to actual features on the ground (such as wetlands, developed areas, coniferous forests, etc.). The unsupervised clustering is a kind of clustering which takes place with minimum input from the operator; no training sample is available and part of the feature space is achieved by identifying natural groupings of the given data. In unsupervised clustering technique, an individual pixel is compared to each cluster to see the closest pixel. Finally, a map of all pixels in the image, classified as to different clusters, each pixel is most likely to belong, is produced. This then must be interpreted by the user as to what the colour patterns may mean in terms of classes, etc. that are actually present in the real-world scene; this requires some knowledge of the scene's feature or cluster content from general experience or personal familiarity with the area imaged. The objective here is to group multi-band spectral response patterns into clusters that are statistically separable. The cluster numbers are initially set and the total number can be varied arbitrarily. There are three unsupervised classification techniques namely K-means clustering,

PCA based K-means clustering and Fuzzy C-Means clustering (FCM). K- means clustering: K-means algorithm is an iterative algorithm that triesto partition the dataset into Kpre-defined distinct non-overlapping subgroups (clusters) where each data point belongs to only one group. It tries to make the inter-cluster data points as similar as possible while also keeping the clusters as different (far) as possible. It assigns data points to a cluster such that the sum of the squared distance between the data points and the cluster's centroid (arithmetic mean of all the data points that belong to that cluster) is at the minimum. The less variation we have within clusters, the more homogeneous (similar) the data points are within the same cluster. The way kmeans algorithm works is as follows:

1. Specify number of clusters K.
2. Initialize centroids by first shuffling the dataset and then randomly selecting K data pointsfor the centroids without replacement.
3. Keep iterating until there is no change to the centroids. i.e assignment of data points to clusters isn't changing.
   - Compute the sum of the squared distance between data points and all centroids.
   - Assign each data point to the closest cluster (centroid).
   - Compute the centroids for the clusters by taking the average of the all-data points thatbelong to each cluster.

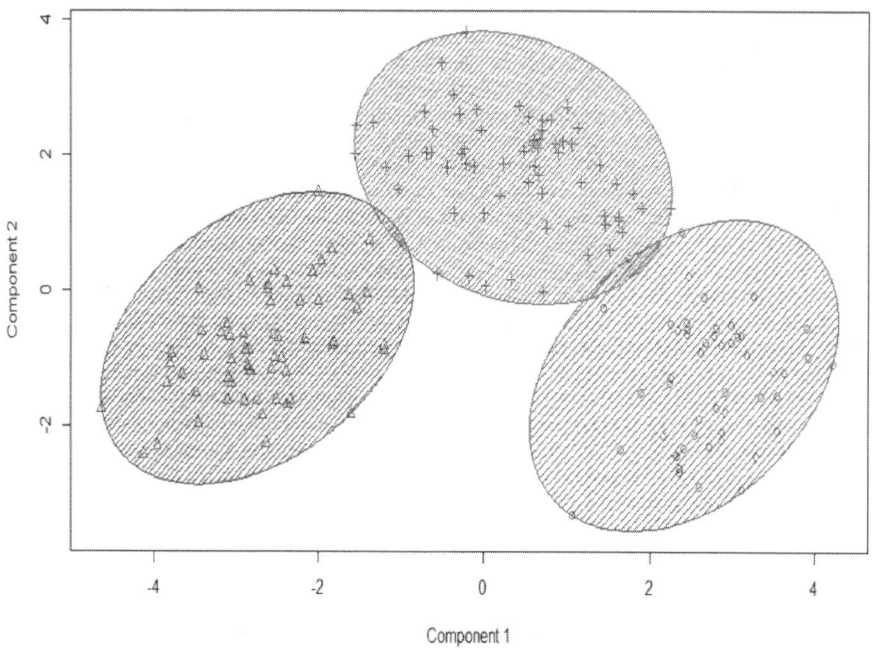

Figure 2.11: K mean clustering

The approach kmeans follows to solve the problem is called Expectation-Maximization. The E-step is assigning the data points to the closest cluster. The M-step is computing the centroidof each cluster. Below is a breakdown of how we can solve it mathematically.

Unsupervised classification finds spectral classes (or clusters) in a multiband image without the analyst's intervention. The Image Classification toolbar aids in unsupervised classification by providing access to the tools to create the clusters, capability to analyze the quality of the clusters, and access to classification tools. Unsupervised classification is a method which examines a large number of unknown pixels and divides into a number of classed based on natural groupings present in the image values. Unlike supervised classification, unsupervised classification does not require analyst-specified training data. The basic premise is that values within a given cover type should be close together in the measurement space (i.e., have similar gray levels), whereas data in different classes should be comparatively well separated (i.e. have very different gray levels) (PCI, 1997; Lillesand and Kiefer, 1994; Eastman, 1995).

The classes that result from unsupervised classification are spectral classed which based on natural groupings of the image values, the identity of the spectral class will not be initial ly known, must compare classified data to some form of reference data (such as larger scale imagery, maps, or site visits) to determine the identity and informational values of the spectral classes. Thus, in the supervised approach, to define useful information categories andthen examine their spectral separability; in the unsupervised approach the computer determines spectrally separable class, and then define their information value. (PCI, 1997; Lillesand and Kiefer, 1994). Unsupervised classification is becoming increasingly popular in agencies involved in long term GIS database maintenance. The reason is that there are now systems that use clustering procedures that are extremely fast and require little in the nature of operational parameters. Thus, it is becoming possible to train GIS analysis with only a general familiarity with remote sensing to undertake classifications that meet typical map accuracy standards. With suitable ground truth accuracy assessment procedures, this tool can provide a remarkably rapid means of producing quality land cover data on a continuing basis. When access to domain knowledge or the experience of an analyst is missing, the data canstill be analyzed by numerical exploration, whereby the data are grouped into subsets or clusters based on statistical similarity. Using the Image Classification toolbar and Training Sample Manager, it was determined the training samples were representative for the area and statistically separate. Therefore, a maximum likelihood classification was performed from thetoolbar.

Change detection

The Land Use and Land Cover (LULC) data files describe the vegetation, water, natural surface, and cultural features on the land surface. The terms land use and land cover areoften used interchangeably, but each term has its own unique meaning. Land cover refers to the surface cover on the ground like vegetation, urban infrastructure, water, bare soil etc. Identification of land cover establishes the baseline information for activities like thematic mapping and change detection analysis. Land use refers to the purpose the land serves, for example, recreation, wildlife habitat, or agriculture. When used together with the phrase Land Use / Land Cover (LULC) generally refers to the categorization or classification of human activities and natural elements on the landscape within a specific time frame based onestablished scientific and statistical methods of analysis of appropriate source materials.

Figure 2.12: Land cover change detection

Land cover relates to the physical aspects of the real world, such as flora, settlements, water bodies, forests, and agricultural regions. Land use relates to how much of a settlement's territory is used for people's dwelling, or how land is used by people (Chaudhary et al., 2008). LULC has become a critical component in measuring and monitoring environmental changes as well as managing natural resources (Kaul and Sopan, 2012). Data from Earth sensing satellites has been increasingly important in mapping the Earth's features and infrastructures, managing natural resources, and analysing environmental change in recent years (Zubair, 2006). Land use land cover (LULC) change detection based on remote sensingdata is an important source of information for various decision support systems. Land conservation, sustainable development, and water resource management all benefit from data produced from land use and land cover change detection (Tewabe and Fentahun, 2020). Remote sensing (RS) and Geographic Information Systems (GIS) are powerful and cost- effective methods for analysing the spatial and temporal change of LULC (Herold et al., 2003; Serra et al., 2008).

LULC mapping in India

Land Use Land Cover (maps of an area provide information to help users to understand the current landscape Annual LULC information on national spatial databases will enable the monitoring of temporal dynamics of agricultural ecosystems, forest conversions, surface water bodies, etc on annual basis. Annual Land use Land cover mapping is carried out at 1 250000 scale and net sown area is estimated apart from generation of the Landuse Land cover information Land use land cover information is generated once in 5 years at 1 50000 scale Land use Land cover is generated at 1 10000 scale which is useful for water land resources planning at village taluk level. Wasteland mapping and monitoring has been carried out at 1:50000 and 1:25000 scales Land use land cover information generated at 1:10000 scale can be used for planning Gram sadak Yojna project.

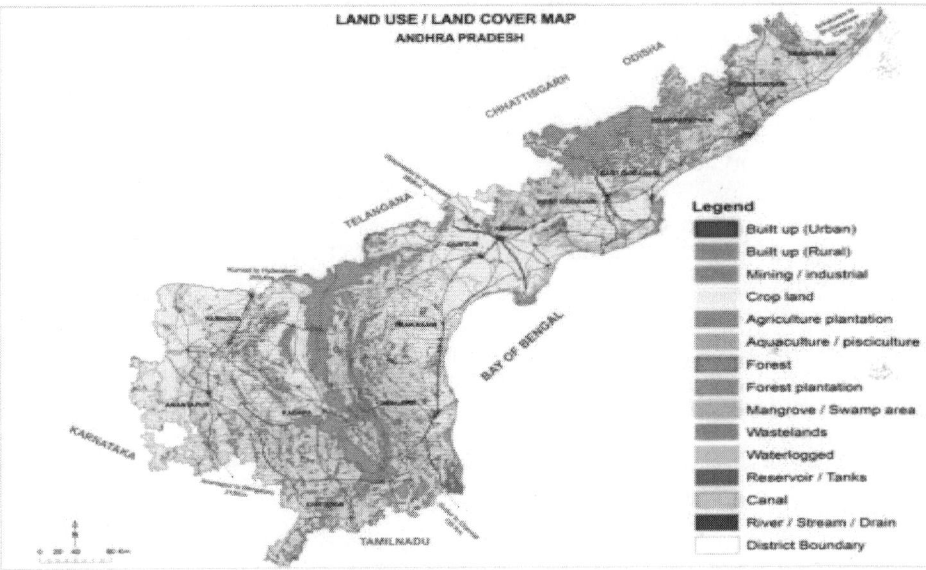

Figure 2.13: Land use land cover map of Andhra Pradesh

Why do we need LULC Maps?

The growth of a society totally depends on its social and economic development. This is the basic reason why socio-economic surveys are carried out. This type of survey includes both spatial and non-spatial datasets. LULC maps play a significant and prime role in planning, management and monitoring programs at local, regional and national levels. This type of information, on one hand, provides a better understanding of land utilization aspects and on the other hand, it plays an important role in the formation of policies and programme required for development planning. For ensuring sustainable development, it is necessary to monitor the ongoing process on land use/land cover pattern over a period of time. In order to achieve sustainable urban development and to check the haphazard development of towns and cities, it is necessary that authorities associated with the urban development generate such planning models so that every bit of available land can be used in most rational and optimal way. This requires the present and past land use/land cover information of the area. LULC maps also help us to study the changes that are happening in our ecosystem and environment. If we have an inch-by-inch information about Land Use/Land Cover of the study unit we can make policies and launch programmes to save our environment.

Applications of LULC maps

- Natural resource management
- Wildlife habitat protection
- Baseline mapping for GIS input
- Urban expansion / encroachment
- Routing and logistics planning for seismic / exploration/resource extraction activities
- Damage delineation (tornadoes, flooding, volcanic, seismic, fire)
- Legal boundaries for tax and property evaluation
- Target detection - identification of landing strips, roads, clearings, bridges, land/water interface

Following are some of the LULC types and their respective classes

Urban or Built-up Land	Residential Commercial and Services Industrial Communications and Utilities Mixed Urban or Built-up Land Other Urban or Built-up Land
Agricultural Land	Cropland and Pasture Orchards, Groves, Vineyards, Nurseries, and Ornamental Horticultural Areas Confined Feeding Operations
Rangeland	Herbaceous Rangeland Shrub and Brush Rangeland Mixed Rangeland
Forest Land	Deciduous Forest Land Evergreen Forest Land Mixed Forest Land
Water	Rivers Streams and Canals Lakes Reservoirs Bays and Estuaries

Wetland	Forested Wetland No forested Wetland
Barren Land	Dry Salt Flats Beaches Sandy Areas Other than Beaches Bare Exposed Rock Strip Mines, Quarries, and Gravel Pits Transitional Areas Mixed Barren Land
Perennial Snow or Ice	Perennial Snowfields Glaciers

Conclusion

Methods of spatial analysis are often used to produce new information from geographic data and there are several spatial analysis techniques available, ranging from simple to complex. GIS and remote sensing offer many capabilities for displaying data at differing scales and based on various attributes and spatial analysis is a source of information from a GIS and is defined by any set of methods whose results change when the locations of the objects being analyzed change. Six types of spatial analysis are queries and reasoning, measurements, transformations, descriptive summaries, optimization, and hypothesis testing. Uncertainty and errors happen in spatial analysis at every stage and utmost care has to be taken tominimize this.

References

1. Grekousis, G. (2020). *Spatial analysis methods and practice: describe–explore–explainthrough GIS*. Cambridge University Press.
2. Fotheringham, S., & Rogerson, P. (Eds.). (2013). *Spatial analysis and GIS*. Crc Press.
3. Paramasivam, C. R., & Venkatramanan, S. (2019). An introduction to various spatial analysis techniques. *GIS and geostatistical techniques for groundwater science*, 23-30.
4. Openshaw, S., & Clarke, G. (2019). Developing spatial analysis functions relevant to GIS environments. In *Spatial analytical perspectives on GIS* (pp. 21-38).
5. Fischer, M. M., Scholten, H. J., & Unwin, D. (2019). Geographic information systems, spatial data analysis and spatial modelling: an introduction. In *Spatial analytical perspectives on GIS* (pp. 3-20). Routledge.
6. Ding, Y., & Fotheringham, A. S. (1992). The integration of spatial analysis and GIS.
7. *Computers, environment and urban systems*, 16(1), 3-19. Lu, D. and Weng, Q., 2007. A survey of image classification methods and techniques for improving

classification performance. *International journal of Remote sensing*, *28*(5), pp.823-870.\

8. Blaschke T, 2010. Object-based image analysis for remote sensing. ISPRS Journal of Photogrammetry and Remote Sensing 65 (2010) 2–16.
9. Object-Based Classification vs Pixel-Based Classification: Comparative Importance of Multi-Resolution Imagery (Robert C. Weih, Jr. and Norman D. Riggan, Jr.)
10. Multiresolution Segmentation: an optimization approach for high-quality multi-scale image segmentation (Martin Baatz and Arno Schape).
11. Trimble eCognition Developer: http://www.ecognition.com
12. Basanna, R. and Wodeyar, A.K., 2013. Supervised classification for LULC change analysis. International Journal of Computer Applications, 66(21).

CHAPTER 3

MACHINE LEARNING AND IT'S APPLICATION

Vinod Kumar Yadav[1], Chandani Dave[1] and Mohammed Meharoof[2]

1 ICAR-Central Institute of fisheries Education, Mumbai-400061
2 Assistant Professor (o/c), College of Fisheries, Payyanur, Kannur-670307

Machine learning

Introduction:

Machine learning is a branch of artificial intelligence (AI) and computer science which focuses on the use of data and algorithms to imitate the way that humans learn,gradually improving its accuracy. Machine learning research enables machines to learn new abilities, gain new knowledge, and reorganize previously acquired knowledge Machine learning is an important component of the growing field of data science.

Through the use of statistical methods, algorithms are trained to make classifications or predictions, and to uncover key insights in data mining projects. These insights subsequently drive decision making within applications and businesses, ideally impacting key growth metrics. As big data continues to expand and grow, the market demand for datascientists will increase. They will be required to help identify the most relevant business questions and the data to answer them. Classical machine learning is often categorized by how an algorithm learns to become more accurate in its predictions.

There are four basic approaches: supervised learning, unsupervised learning, semi-supervised learning and reinforcement learning. The type of algorithm data scientists choose to use depends on what type of data they want to predict. Decision tree, NaiveBayes, Support vector machine, artificial neural network,

K nearest neighbor, Deep learning, Ensemble learning are some examples of machine learning

Why Machine Learning?

Most problems in Remote Sensing require 'detecting' objects or land cover types from images. Earth Observation data is noisy. Hard to determine rules that can separate one class from the other. Building models require expert knowledge and can be tedious. Machine learning algorithms takes the guesswork out of determining thresholds and cut-offs. Machine learning is growing in importance due to increasingly enormous volumes and variety of data, the access and affordability of computational power, and the availability of high speed Internet. These digital transformation factors make it possible for one to rapidly and automatically develop models that can quickly and accurately analyze extraordinarily large and complex data sets. There are a multitudeof use cases that machine learning can be applied in order to cut costs, mitigate risks, and improve overall quality of life including recommending products/services, detecting cybersecurity breaches, and enabling self-driving cars. With greater access to data and computation power, machine learning is becoming more ubiquitous every day and will soon be integrated into many facets of human life.

Supervised vs. Unsupervised Learning:

These two are the broad categories of machine learning model approach

Supervised Learning	Unsupervised Learning
Supervised Learning can be used for 2 different types of problems i.e. regression and classification	Unsupervised Learning can be used for 2 different types of problems i.e. clustering and association.
Input Data is provided to the model along with the output in the Supervised Learning.	Only input data is provided in Unsupervised Learning.
Output is predicted by the Supervised Learning.	Hidden patterns in the data can be found using the unsupervised learning model.
Labeled data is used to train supervised learning algorithms.	Unlabeled data is used to train unsupervised learning algorithms.
Accurate results are produced using a supervised learning model.	The accuracy of results produced are less in unsupervised learning models.
Training the model to predict output when a new data is provided is the objective of Supervised Learning.	Finding useful insights, hidden patterns from the unknown dataset is objective of unsupervised learning.
Supervised Learning includes various algorithms such as Bayesian Logic, Decision Tree, Logistic Regression, Linear Regression, Multi-class Classification, SVM	Unsupervised Learning includes various algorithms like KNN, Apriori Algorithm, and Clustering.

In Supervised Learning, for right prediction of output, the model has to be trained for each data, hence Supervised Learning does not have close resemblance to AI.	Unsupervised Learning has more resemblance to Artificial Intelligence, as it keeps learning new things with more experience.
In scenarios where one is aware of output and input data, supervised learning can be used.	In the scenarios where one is not aware of output data, but is only aware of the input data then Unsupervised Learning used.
Computational Complexity is very complex in Supervised Learning compared to Unsupervised Learning.	There is less computational complexity when compared to Supervised Learning.
Some of the applications of Supervised Learning are handwriting detection, pattern recognition, speech recognition etc.	Some of the applications of Unsupervised Learning are detecting fraudulent transactions, data preprocessing etc.

Supervised learning

Supervised learning, as the name indicates, has the presence of a supervisor as a teacher. Basically supervised learning is when we teach or train the machine using data that is well- labelled. Which means some data is already tagged with the correct answer. After that, the machine is provided with a new set of examples (data) so that the supervised learning algorithm analyses the training data(set of training examples) and produces a correct outcome from labeled data.

Supervised learning is classified into two categories of algorithms:

Classification: A classification problem is when the output variable is a category, such as

-Red or –blue, –disease or –no disease.

Regression: A regression problem is when the output variable is a real value, such as –dollars ll or –weight ll .

Supervised learning deals with or learns with —labeledII data. This implies that some data is already tagged with the correct answer.

Types:-

- Regression
- Logistic Regression
- Classification
- Naive Bayes Classifiers
- K-NN (k nearest neighbors)
- Decision Trees
- Support Vector Machine

Unsupervised learning

Unsupervised learning is the training of a machine using information that is neither classifiednor labeled and allowing the algorithm to act on that information without guidance. Here the task of the machine is to group unsorted information according to similarities, patterns, and differences without any prior training of data. Unlike supervised learning, no teacher is provided that means no training will be given to the machine. Therefore the machine is restricted to find the hidden structure in unlabeled data by itself.

Unsupervised learning is classified into two categories of algorithms:

Clustering: A clustering problem is where you want to discover the inherent groupings in the data, such as grouping customers by purchasing behavior.

Association: An association rule learning problem is where you want to discover rules that describe large portions of your data, such as people that buy X also tend to buy Y.

Types:-

- Hierarchical clustering
- K-means clustering
- Principal Component Analysis

Classification

Classification is a supervised machine learning method where the model tries to predict the correct label of a given input data (Sen et al., 2022). In classification, the model is fully trained using the training data, and then it is evaluated on test data before being used to perform prediction on new unseen data.

Why do we need to do classification?

For many applications, it is important to know what type of land cover pixel represents. (i.e.water)

For a given land cover, we often need to know about different types of land uses (i.e. agriculture vs. forest)

We also need to quantify land use patterns (i.e. how much agricultural area in a region) For detecting changes, we need to compare how a pixel changed from Class A to Class B

The main classification problems include:

1. Binary classification
2. Multiple class classification
3. Multiple label classification

Binary classification: Binary is a type of problem in classification in machine learning that has only two possible outcomes (Karaca, 2012). For example, yes or no, true or false, spam or not spam, etc. Examples include: Logistic Regression, Decision Trees, Naïve Bayes

Multiple class classification: Multi-class is a type of classification problem with more than two outcomes and does not have the concept of normal and abnormal outcomes. Here each outcome is assigned to only one label. For example, classifying images, classifying species, and categorizing faces, among others.

Examples include: decision Tree, Random Forest, k-Nearest Neighbors, Naive Bayes, Artificial neural network (ANN)

Multiple label classification: Multi-label is a type of classification problem that may have more than one class label assigned to the data. Here the model will have multiple outcomes. For example, a book or a movie can be categorized into multiple genres, or an image can have multiple objects.

Examples: Multiple label Gradient Boosting, Multiple label Random Forests, Using different classification algorithms for each class label

The general workflow for classification is

- Collect training data.

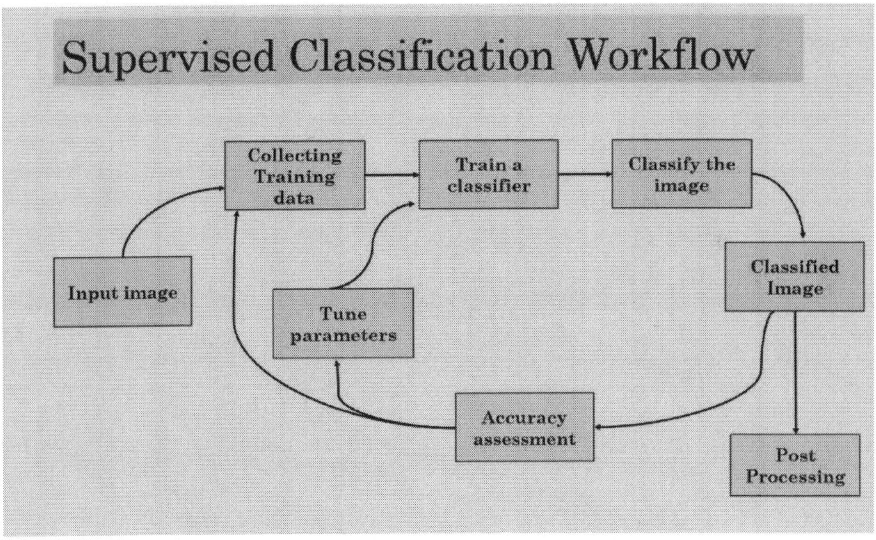

Figure 3.1 Workflow of Supervised classification

- Assemble features which have a property that stores the known class label and propertiesstoring numeric values for the predictors.
- Instantiate a classifier.
- Set its parameters if necessary.
- Train the classifier using the training data.
- Classify an image or feature collection.Estimate classification error with independent validation data.

Different supervised ML algorithms:

- Support vector machine
- Decision Tree
- Neural network
- KNN (K Nearest Neighbour)

Support Vector Machines

Support Vector Machine or SVM is one of the most popular Supervised Learning algorithms, which is used for Classification as well as Regression problems (Nasteski, 2017). However, primarily, it is used for Classification problems in Machine Learning. The goal of the SVM algorithm is to create the best line or decision boundary that can segregate n-dimensional space into classes so that we can easily put the new data point in the correct category in the future. This best decision boundary is called a hyperplane. SVM chooses the extreme points/vectors that help in creating the hyperplane. These extreme cases are called as support vectors, and hence algorithm is termed as Support Vector Machine. Consider the below diagram in which there are two different categories that are classified using a decision boundary or hyperplane:

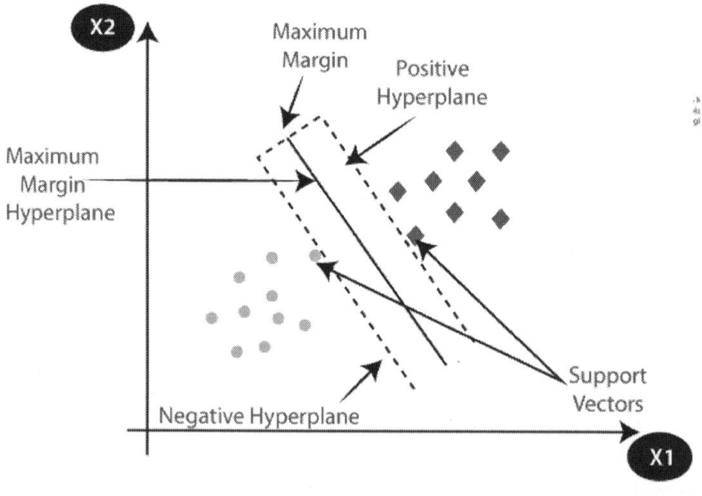

Figure 3.2 Classification of different categories using hyperplane

Example: SVM can be understood with the example that we have used in the KNN classifier.Suppose we see a strange cat that also has some features of dogs, so if we want a model that can accurately identify whether it is a cat or dog, so such a model can be created by using the SVM algorithm. We will first train our model with lots of images of cats and dogs so that it can learn about different features of cats and dogs, and then we test it with this strange creature. So as support vector creates a decision boundary between these two data (cat and dog) and choose extreme cases (support vectors), it will see the extreme case of cat and dog. On the basis of the support vectors, it will classify it as a cat. Consider the figure 3.3:

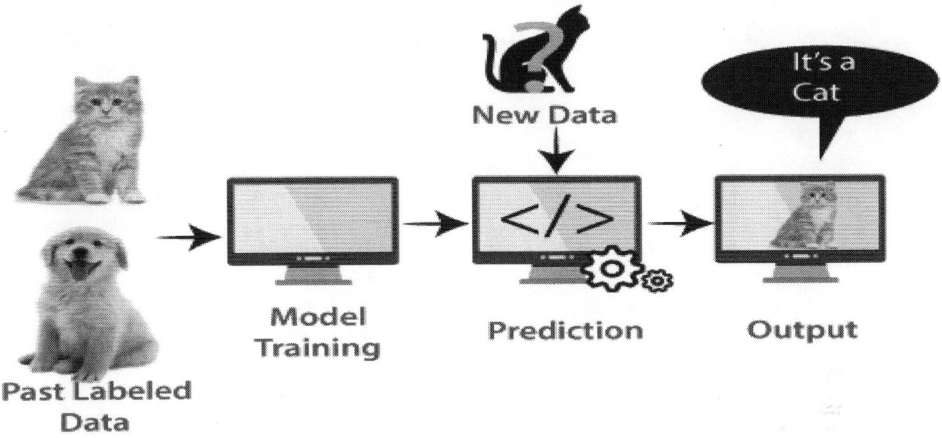

Figure 3.3 : Example of animal identification using by using the SVM algorithm

SVM algorithm can be used for Face detection, image classification, text categorization, etc.Types of SVM:

SVM can be of two types:

Linear SVM: Linear SVM is used for linearly separable data, which means if a dataset can be classified into two classes by using a single straight line, then such data is termed as linearly separable data, and classifier is used called as Linear SVM classifier.

Non-linear SVM: Non-Linear SVM is used for non-linearly separated data, which means if a dataset cannot be classified by using a straight line, then such data is termed as non-linear data and classifier used is called as Non-linear SVM classifier.

Hyperplane and Support Vectors in the SVM algorithm:

Hyperplane: There can be multiple lines/decision boundaries to segregate the classes in n- dimensional space, but we need to find out the best decision boundary that helps to classify the data points. This best boundary is known as the hyperplane of SVM.

The dimensions of the hyperplane depend on the features present in the dataset, which meansif there are 2 features (as shown in image), then hyperplane will be a straight line. And if there are 3 features, then hyperplane will be a 2-dimension plane. Always create a hyperplane that has a maximum margin, which means the maximum distance between the data points.

Support Vectors: The data points or vectors that are the closest to the hyperplane and which affect the position of the hyperplane are termed as Support Vector. Since these vectors support the hyperplane, hence called a Support vector.

How does SVM works? Linear SVM

The working of the SVM algorithm can be understood by using an example Figure 3.4 and Figure 3.5. Suppose we have a dataset that has two tags (green and blue), and the dataset has two features x1 and x2. We want a classifier that can classify the pair(x1, x2) of coordinates in either green or blue. Consider the below image:

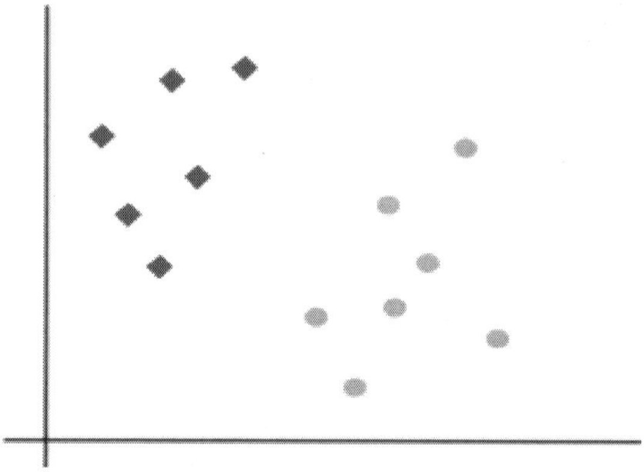

Figure 3.4: Datasets to be classified

So as it is 2-d space so by just using a straight line, we can easily separate these two classes. But there can be multiple lines that can separate these classes.

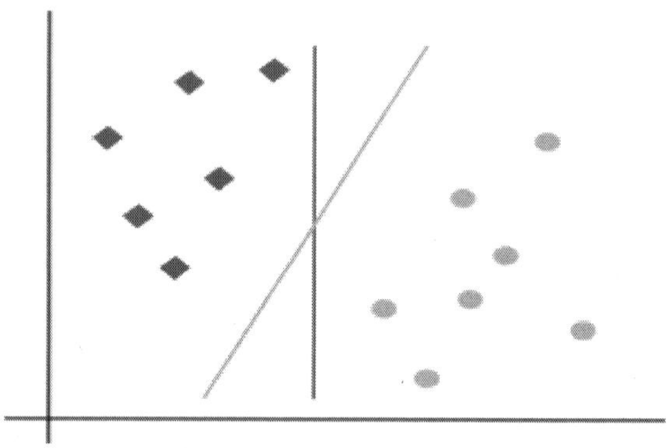

Figure 3.5: possible ways to separate the classes

Hence, the SVM algorithm helps to find the best line or decision boundary; this best boundary or region is called as a hyperplane (Figure 3.6). SVM algorithm finds the closest point of the lines from both the classes. These points are called support vectors. The distance between thevectors and the hyperplane is called as margin. And the goal of SVM is to maximize this margin. The hyperplane with maximum margin is called the optimal hyperplane.

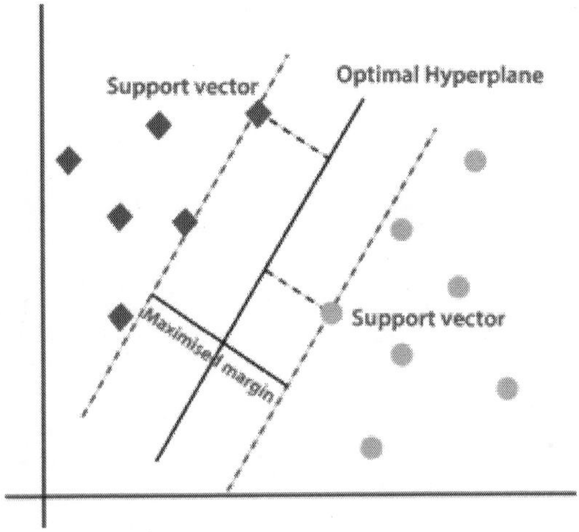

Figure 3.6: Optimal hyperplane that can separate the vectors with the maximum margin

Non-Linear SVM:

If data is linearly arranged, then we can separate it by using a straight line, but for non-linear data, we cannot draw a single straight line so to separate these data points, we need to add one more dimension. For linear data, wehave used two dimensions x and y, so for non-linear data, we will add a third dimension z. It can be calculated as: z=x2 +y2

By adding the third dimension, the sample space will become as below image:

(A) **(B)**

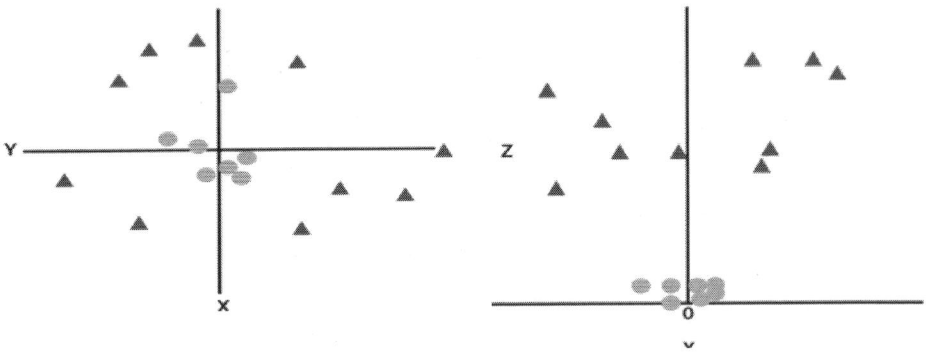

Figure 3.7: Representation of non-linear data in two dimension (A) and three dimension (B)

So now, SVM will divide the datasets into classes in the following way. Consider the below image:

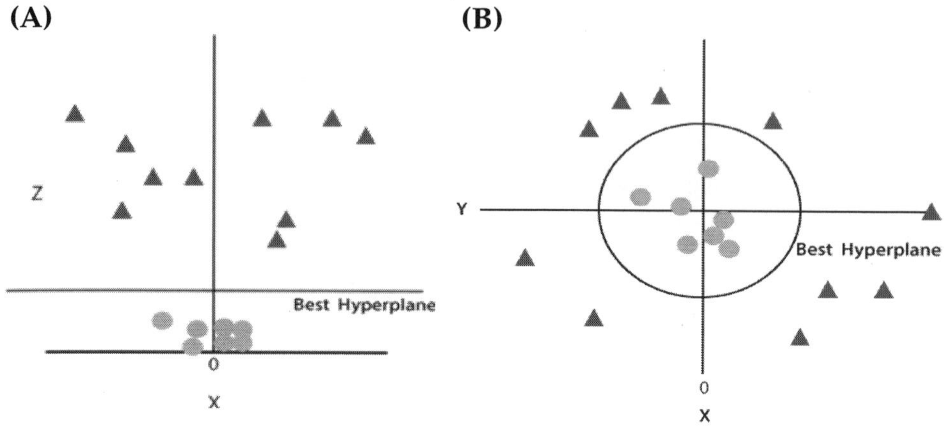

Figure: 3.8 Optimal hyperplane in 3-D (A) and 2-D (B)

Since we are in 3-d Space, hence it is looking like a plane parallel to the x-axis. If we convertit in 2d space with z=1, then it will become as Figure 3.8 (A and B). Hence we get a circumference of radius 1 in case of non-linear data.

Bias and Variance in Machine Learning:

Bias: Bias refers to the error due to overly simplistic assumptions in the learning algorithm. These assumptions make the model easier to comprehend and learn but might not capture theunderlying complexities of the data. It is the error due to the model's inability to representthe true relationship between input and output accurately. When a model has poor performance both on the training and testing data means high bias because of the simple model, indicating underfitting.

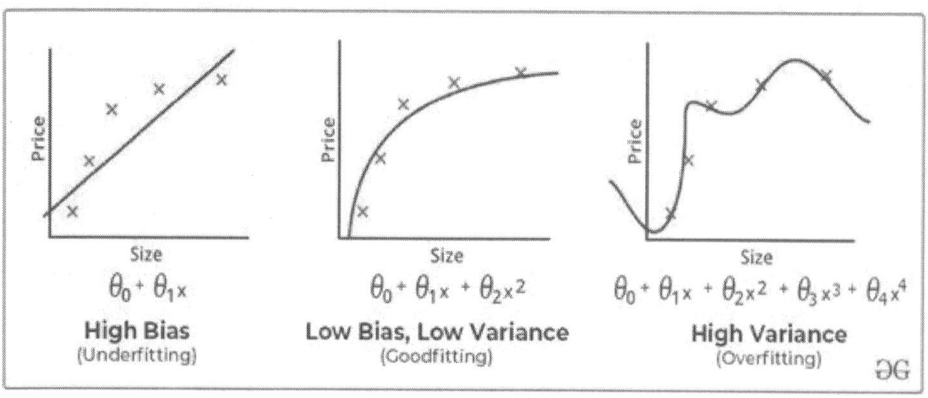

Figure 3.9: Bias and Variance

Variance: Variance, on the other hand, is the error due to the model's sensitivity to fluctuations in the training data. It's the variability of the model's predictions for different instances of training data. High variance occurs when a model learns the training data's noise and random fluctuations rather than the underlying pattern. As a result, the model performs well on the training data but poorly on the testing data, indicating overfitting.

Underfitting in Machine Learning:

A statistical model or a machine learning algorithm is said to have underfitting when a modelis too simple to capture data complexities. It represents the inability of the model to learn the training data effectively result in poor performance both on the training and testing data. In simple terms, an underfit model's are inaccurate, especially when applied to new, unseen examples. It mainly happens when we uses very simple model with overly simplified assumptions. To address underfitting problem of the model, we need to use more complex models, with enhanced feature representation, and less regularization.

Overfitting in Machine Learning:

A statistical model is said to be overfitted when the model does not make accurate predictions on testing data. When a model gets trained with so much data, it starts learning from the noise and inaccurate data entries in our data set and when testing with test data results in High variance (Xiao, et al., 2015). Then the model does not categorize the data correctly, because of too many details and noise. The causes of overfitting are the non-parametric and non-linear methods because these types of machine learning algorithms have more freedom in building the model based on the dataset and therefore they can really build unrealistic models. A solution to avoid overfitting is using a linear algorithm if we have linear data or using the parameters like the maximal depth if we are using decision trees.

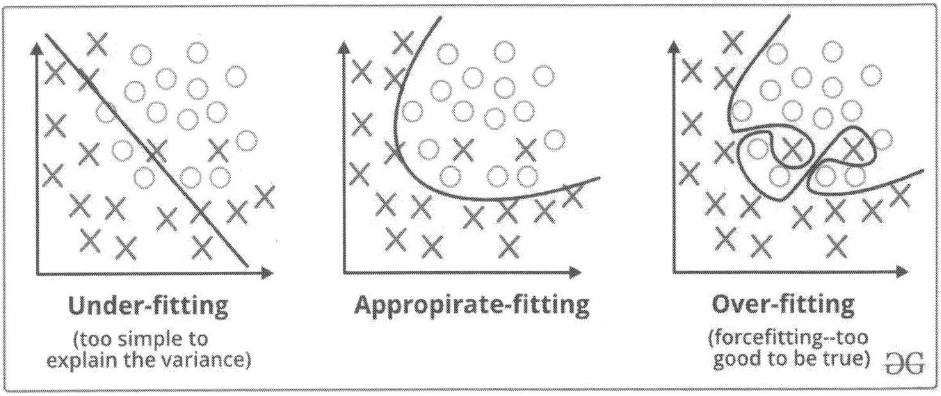

Figure 3.10: Under-fitting, appropriate-fitting and over-fitting in machine learning

Good Fit in a Statistical Model

Ideally, the case when the model makes the predictions with 0 error, is said to have a good fiton the data. This situation is achievable at a spot between overfitting and underfitting. In order to understand it, we will have to look at the performance of our model with the passage of time, while it is learning from the training dataset. With the passage of time, our model will keep on learning, and thus the error for the model on the training and testing data will keep on decreasing. If it will learn for too long, the model will become more prone to overfitting due to the presence of noise and less useful details (Yieng, 2018). Hence the performance of our model will decrease. In order to get a good fit, we will stop at a point just before where the error starts increasing. At this point, the model is said to have good skills in training datasets as well as our unseen testing dataset.

Decision Trees:

What is a Decision Tree?

A decision tree is a predictive model that uses a flowchart-like structure to make decisions based on input data. It divides data into branches and assigns outcomes to leaf nodes. Decision trees are used for classification and regression tasks, providing easy-to-understand models. A decision tree is a hierarchical model used in decision support that depicts decisions and their potential outcomes, incorporating chance events, resource expenses, and utility. This algorithmic model utilizes conditional control statements and is non-parametric, supervised learning, useful for both classification and regression tasks. The tree structure is comprised of a root node, branches, internal nodes, and leaf nodes, forming a hierarchical, tree-like structure. It is a tool that has applications spanning several different areas. Decision trees can be used for classification as well as regression problems. The name itself suggests that it uses a flowchart like a tree structure to show the predictions that result from a series of feature-based splits (Figure 3.11). It starts with a root node and ends with a decision made by leaves.

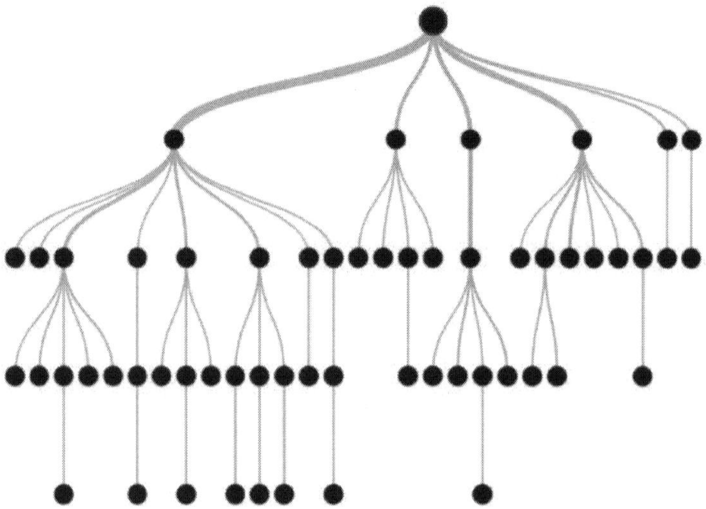

Figure 3.11: Decision tree

Importance of decision tree in machine learning

Decision trees in machine learning provide an effective method for making decisions because they lay out the problem and all the possible outcomes. It enables developers to analyze the possible consequences of a decision, and as an algorithm accesses more data, it can predict outcomes for future data.

Types of decision trees in machine learning

Decision trees in machine learning can either be classification trees or regression trees. Together, both types of algorithms fall into a category of -classification and regression trees ‖ and are sometimes referred to as CART. Their respective roles are to -classify ‖ and to -predict. ‖

1. Classification trees:

Classification trees determine whether an event happened or didn't happen. Usually, this involves a -yes ‖ or -no ‖ outcome.

We often use this type of decision-making in the real world. Here are a few examples to help contextualize how decision trees work for classification:

Example 1: How to spend your free time after work

What you do after work in your free time can be dependent on the weather. If it is sunny, youmight choose between having a picnic with a friend, grabbing a drink with a colleague, or running errands. If it is raining, you might opt to stay home and watch a movie instead. There is a clear outcome. In this case, that is classified as whether to -go out ‖ or -stay in. ‖

Example 2: Homeownership based on age and income

In a classification tree, the data set splits according to its variables. There are two variables, age and income, that determine whether or not someone buys a house. If training data tells usthat 70 percent of people over age 30 bought a house, then the data gets split there, with age becoming the first node in the tree. This split makes the data 80 percent -pure. ‖ The second node then addresses income from there. If you want to get started on understanding how decision trees work in machine learning, consider registering for these guided projects to apply your skills to real-world projects. You can complete them in two hours or less:

2. Regression trees:

Regression trees, on the other hand, predict continuous values based on previous data or information sources. For example, they can predict the price of gasoline or whether a customer will purchase eggs (including which type of eggs and at which store).

This type of decision-making is more about programming algorithms to predict what is likelyto happen, given previous behavior or trends.

Example 1: Housing prices in Colorado

Regression analysis could be used to predict the price of a house in Colorado, which is plotted on a graph. The regression model can predict housing prices in the coming years using data points of what prices have been in previous years. This relationship is a linear regression since housing prices are expected to continue rising. Machine learning helps us predict specific prices based on a series of variables that have been true in the past.

Example 2: Bachelor's degree graduates in 2025

A regression tree can help a university predict how many bachelor's degree students there will be in 2025. On a graph, one can plot the number of degree-holding students between 2010 and 2022. If the number of university graduates increases linearly each year, then regression analysis can be used to build an algorithm that predicts the number of students in 2025.

To get started on how decision tree algorithms work in predictive machine learning models, take a look at these guided projects. Each project takes less than two hours, and they are based on real-world examples so you can elevate your skills:

Classification and Regression Tree (CART) is a predictive algorithm used in machine learning that generates future predictions based on previous values. These decision trees are at the core of machine learning, and serve as a basis for other machine learning algorithms such as random forest, bagged decision trees, and boosted decision trees.

How does the Decision Tree algorithm Work?

In a decision tree, for predicting the class of the given dataset, the algorithm starts from the root node of the tree. This algorithm compares the values of root attribute with the record (real dataset) attribute and, based on the comparison, follows the branch and jumps to the next node. For the next node, the algorithm again compares the attribute value with the other sub-nodes and move further. It continues the process until it reaches the leaf node of the tree.

The complete process can be better understood using the below algorithm:

Step-1: Begin the tree with the root node, says S, which contains the complete dataset.

Step-2: Find the best attribute in the dataset using Attribute Selection Measure (ASM).

Step-3: Divide the S into subsets that contains possible values for the best attributes.

Step-4: Generate the decision tree node, which contains the best attribute.

Step-5: Recursively make new decision trees using the subsets of the dataset created in step -

3. Continue this process until a stage is reached where you cannot further classify the nodes and called the final node as a leaf node.

Decision Tree Terminologies:

Root Nodes – It is the node present at the beginning of a decision tree from this node the population starts dividing according to various features.

Decision Nodes – the nodes we get after splitting the root nodes are called Decision Node

Leaf Nodes – the nodes where further splitting is not possible are called leaf nodes or terminal nodes

Sub-tree – just like a small portion of a graph is called sub-graph similarly a sub-section of this decision tree is called sub-tree.

Pruning – is nothing but cutting down some nodes to stop overfitting.

Different terminologies are depicted in figure 3.12 for better understanding.

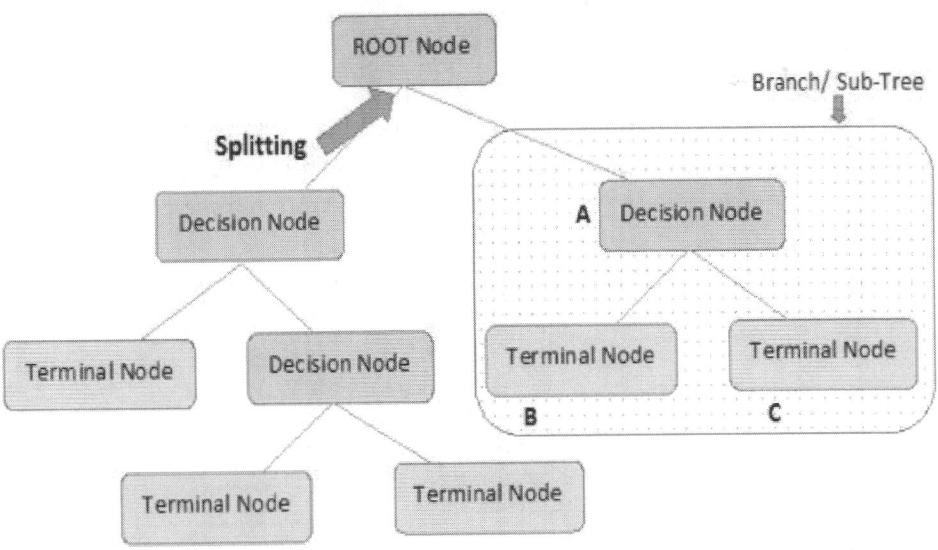

Figure 3.12: Decision tree and its different nodes

Example of Decision Tree:

Let's understand decision trees with the help of an example:

Day	Weather	Temperature	Humidity	Wind	Play?
1	Sunny	Hot	High	Weak	No
2	Cloudy	Hot	High	Weak	Yes
3	Sunny	Mild	Normal	Strong	Yes
4	Cloudy	Mild	High	Strong	Yes
5	Rainy	Mild	High	Strong	No
6	Rainy	Cool	Normal	Strong	No
7	Rainy	Mild	High	Weak	Yes
8	Sunny	Hot	High	Strong	No
9	Cloudy	Hot	Normal	Weak	Yes
10	Rainy	Mild	High	Strong	No

Decision trees are upside down which means the root is at the top and then this root is split into various several nodes. Decision trees are nothing but a bunch of if-else statements in layman terms. It checks if the condition is true and if it is then it goes to the next node attached to that decision.

In the below diagram the tree will first ask what is the weather? Is it sunny, cloudy, or rainy? If yes then it will go to the next feature which is humidity and wind. It will again check if there is a strong wind or weak, if it's a weak wind and it's rainy then the person may go and play.

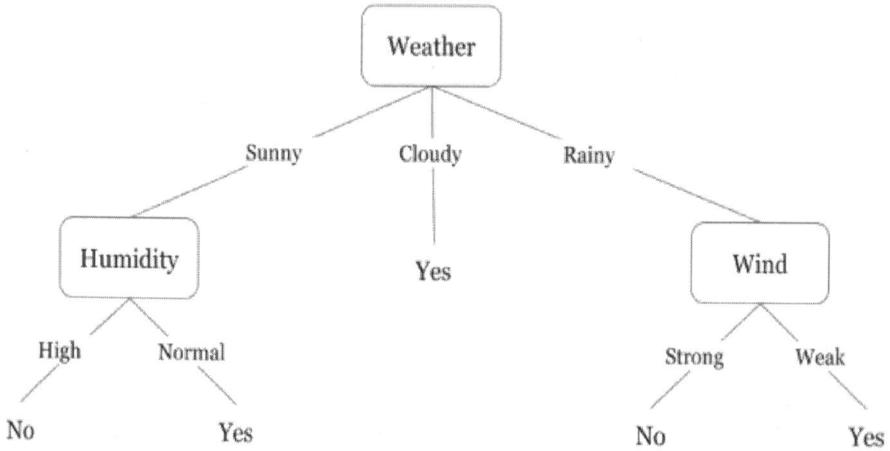

Figure 3.13: Upside down decision tree

Did you notice anything in the above flowchart? We see that if the weather is cloudy then wemust go to play. Why didn't it split more? Why did it stop there?

To answer this question, we need to know about few more concepts like entropy, information gain, and Gini index. But in simple terms, I can say here that the output for the training dataset is always yes for cloudy weather, since there is no disorderliness here we don't need to split the node further. The goal of machine learning is to decrease uncertainty or disorders from the dataset and for this, we use decision trees. Now must think how do I know what should be the root node? what should be the decision node? when should I stop splitting? To decide this, there is a metric called -Entropy || which is the amount of uncertainty in the dataset.

Entropy: Entropy is nothing but the uncertainty in our dataset or measure of disorder. Let me try to explain this with the help of an example.

Suppose you have a group of friends who decides which movie they can watch together on

Sunday. There are 2 choices for movies, one is -Lucy and the second is -Titanic || and now everyone has to tell their choice. After everyone gives their answer we see that -Lucy || gets 4 votes and -Titanic gets 5 votes. Which movie do we watch now? Isn't it hard to choose 1 movie now because the votes for both the movies are somewhat equal.

This is exactly what we call disorderness, there is an equal number of votes for both the movies, and we can't really decide which movie we should watch. It would have been much easier if the votes for -Lucy were 8 and for -Titanic it was 2. Here we could easily say that the majority of votes are for -Lucy || hence everyone will be watching this movie.

In a decision tree, the output is mostly -yes or -no The formula for Entropy is shown below:

$$E(S) = -P_{(+)} logp_{(+)} - p_{(-)} logp_{(-)}$$

Here p+ is the probability of positive classp– is the probability of negative class S is the subset of the training example

How do Decision Trees use Entropy?

Now we know what entropy is and what is its formula, Next, we need to know that how exactly does it work in this algorithm. Entropy basically measures the impurity of a node. Impurity is the degree of randomness; it tells how random our data is. Apure sub-split means that either you should be getting -yes, or you should be getting -no || . Suppose a feature has 8

-yes and 4 -no initially, after the first split the left node gets 5 yes' and 2 no'where as right node gets 3 yes' and 2 no'. We see here the split is not pure, why? Because we can still see some negative classes in both the nodes. In order to make a decision tree, we need to calculate the impurity of each split, and when the purity is 100%, we make it as a leaf node. To check the impurity of feature 2 and feature 3 we will take the help for Entropy formula.

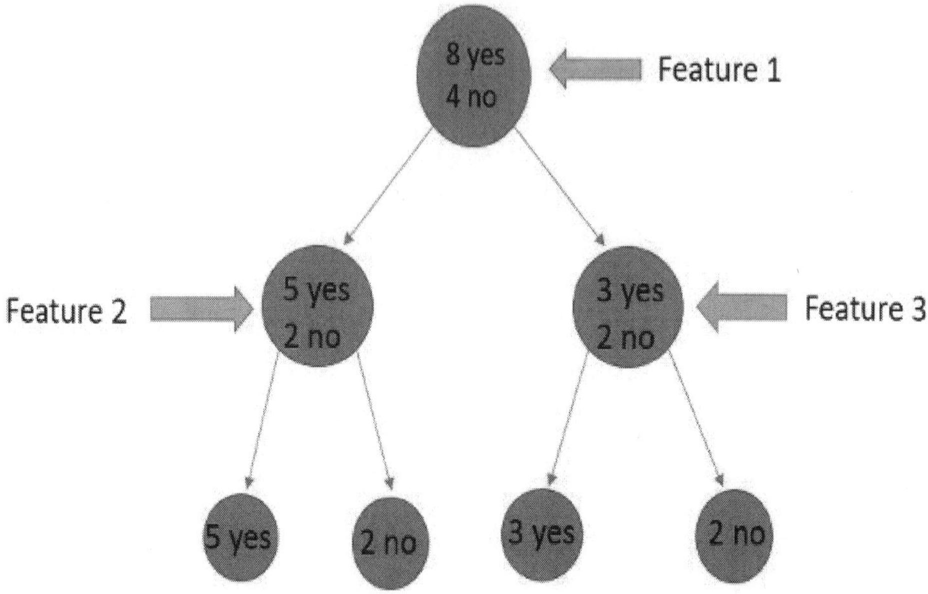

Figure 3.14: Diagram showing the use of entropy in decision tree

$$= -\left(\frac{5}{7}\right) log_2 \left(\frac{5}{7}\right) - \left(\frac{2}{7}\right) log_2 \left(\frac{2}{7}\right)$$

$$= -(0.71^* - 0.49) - (0.28^* - 1.83)$$

$$= -(-0.34) - (-0.51)$$

$$= 0.34 + 0.51$$

$$= 0.85$$

For feature 3,

$$= -\left(\frac{3}{5}\right) log_2 \left(\frac{3}{5}\right) - \left(\frac{2}{5}\right) log_2 \left(\frac{2}{5}\right)$$

$$= -(0.6^* - 0.73) - (0.4^* - 1.32)$$

$$= -(-0.438) - (-0.528)$$

$$= 0.438 + 0.528$$

$$= 0.966$$

We can clearly see from the tree itself that left node has low entropy or more purity than right node since left node has a greater number of -yes and it is easy to decide here.

Always remember that the higher the Entropy, the lower will be the purity and the higher will be the impurity.

As mentioned earlier the goal of machine learning is to decrease the uncertainty or impurity in the dataset, here by using the entropy we are getting the impurity of a particular node, we don't know if the parent entropy or the entropy of a particular node has decreased or not.

For this, we bring a new metric called -Information gain which tells us how much the parent entropy has decreased after splitting it with some feature.

Information Gain

Information gain measures the reduction of uncertainty given some feature and it is also adeciding factor for which attribute should be selected as a decision node or root node.

$$Information\ Gain = E(Y) - E(Y|X)$$

It is just entropy of the full dataset – entropy of the dataset given some feature.

To understand this better let's consider an example: Suppose our entire population has a total of 30 instances. The dataset is to predict whether the person will go to the gym or not. Let's say 16 people go to the gym and 14 people don't.

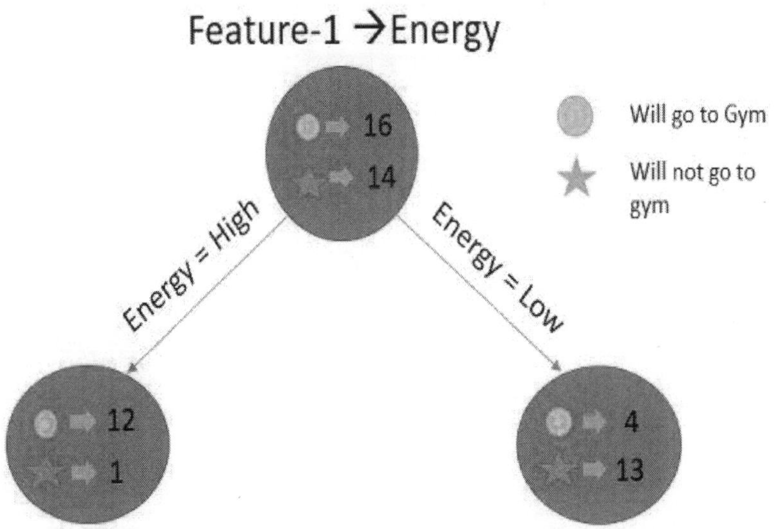

Figure 3.15: considering energy as deciding factor for people will go to gym or not

Now we have two features to predict whether he/she will go to the gym or not. Feature 1 is -Energy which takes two values -high and -low. Feature 2 is -Motivation which takes 3 values -No motivation, -Neutral and -Highly motivated. Let's see how our decision tree will be made using these 2 features. We'll use information gain to decide which feature should be the root node and which feature should be placed after the split.

Let's calculate the entropy

$$E(Parent) = -\left(\frac{16}{30}\right)\log_2\left(\frac{16}{30}\right) - \left(\frac{14}{30}\right)\log_2\left(\frac{14}{30}\right) \approx 0.99$$

$$E(Parent|Energy = \text{"high"}) = -\left(\frac{12}{13}\right)\log_2\left(\frac{12}{13}\right) - \left(\frac{1}{13}\right)\log_2\left(\frac{1}{13}\right) \approx 0.39$$

$$E(Parent|Energy = \text{"low"}) = -\left(\frac{4}{17}\right)\log_2\left(\frac{4}{17}\right) - \left(\frac{13}{17}\right)\log_2\left(\frac{13}{17}\right) \approx 0.79$$

To see the weighted average of entropy of each node we will do as follows:

$$E(Parent|Energy) = \frac{13}{30} * 0.39 + \frac{17}{30} * 0.79 = 0.62$$

Now we have the value of E(Parent) and E(Parent | Energy), information gain will be:

$$\begin{aligned} Information\ Gain &= E(parent) - E(parent|energy) \\ &= 0.99 - 0.62 \\ &= 0.37 \end{aligned}$$

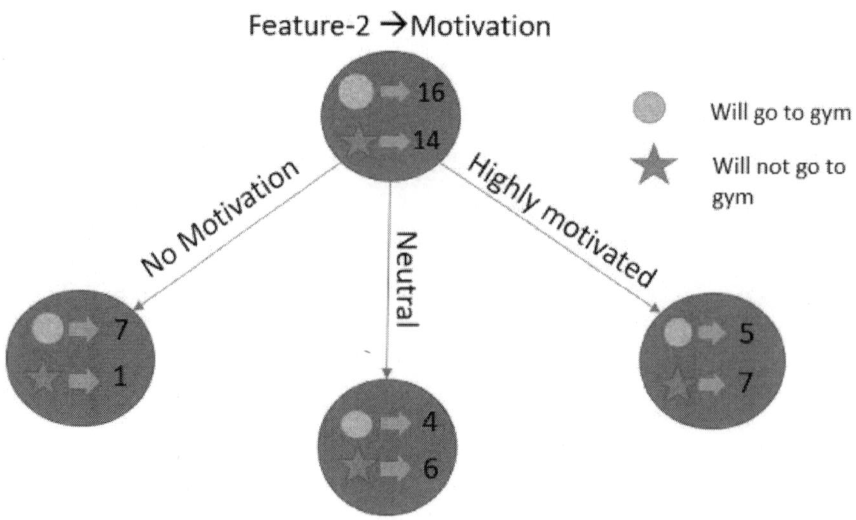

Figure 3.16: considering otivation as deciding factor for people will go to gym or not

Our parent entropy was near 0.99 and after looking at this value of information gain, we can say that the entropy of the dataset will decrease by 0.37 if we make -Energy ll as our root node.

Similarly, we will do this with the other feature -Motivation ll and calculate its information gain.

Let's calculate the entropy here:

$$E(Parent) = 0.99$$

$$E\left(Parent|Motivation = "No\ motivation"\right) = -\left(\frac{7}{8}\right)log_2\left(\frac{7}{8}\right) - \frac{1}{8}\log_2\left(\frac{1}{8}\right) = 0.54$$

$$E(Parent|Motivation = "Neutral") = -\left(\frac{4}{10}\right)log_2\left(\frac{4}{10}\right) - \left(\frac{6}{10}\right)\log_2\left(\frac{6}{10}\right) = 0.97$$

$$E(Parent|Motivation = "Highly\ motivated") = -\left(\frac{5}{12}\right)log_2\left(\frac{5}{12}\right) - \left(\frac{7}{12}\right)\log_2\left(\frac{7}{12}\right) = 0.98$$

To see the weighted average of entropy of each node we will do as follows:

$$E(Parent|Motivation) = \frac{8}{30}*0.54 + \frac{10}{30}*0.97 + \frac{12}{30}*0.98 = 0.86$$

Now we have the value of E(Parent) and E(Parent I Motivation), information gain will be:

$$\begin{aligned} Information\ Gain &= E(Parent) - E(Parent|Motivation) \\ &= 0.99 - 0.86 \\ &= 0.13 \end{aligned}$$

We now see that the -Energy ll feature gives more reduction which is 0.37 than the

Motivation feature. Hence we will select the feature which has the highest information gain and then split the node based on that feature. In this example - Energy will be our root node and we'll do the same for sub-nodes. Here we can see that when the energy is -high ll the entropy is low and hence we can say a person will definitely go to the gym if he has high energy, but what if the energy is low? We will again split the node based on the new feature which is -Motivation.

When to Stop Splitting?

You must be asking this question to yourself that when do we stop growing our tree? Usually, real-world datasets have a large number of features, which will result in a large number of splits, which in turn gives a huge tree. Such trees take time to build and can leadto overfitting. That means the tree will give very good accuracy on the training dataset but will give bad accuracy in test data.

There are many ways to tackle this problem through hyperparameter tuning. We can set the maximum depth of our decision tree using themax_depth parameter. The more the valueof max_depth, the more complex your tree will be. The training error will off-course decrease if we increase the max_depth value but when our test data comes into the picture, we will get a very bad accuracy. Hence you need a value that will not overfit as well as underfit our data and for this, you can use GridSearchCV. Another way is to set the minimum number of samples for each spilt. It is denoted by min_samples_split. Here we specify the minimum number of samples required to do a spilt. For example, we can use a minimum of 10 samples to reach a decision. That means if a node has less than 10 samples then using this parameter, we can stop the further splitting of this node and make it a leaf node.

There are more hyperparameters such as :

min_samples_leaf– represents the minimum number of samples required to be in the leafnode. The more you increase the number, the more is the possibility of overfitting. max_features – it helps us decide what number of features to consider when looking for thebest split.

Pruning:

Pruning is another method that can help us avoid overfitting. It helps in improving the performance of the tree by cutting the nodes or sub-nodes which are not significant. Additionally, it removes the branches which have very low importance.

There are mainly 2 ways for pruning:

Pre-pruning – we can stop growing the tree earlier, which means we can prune/remove/cut a node if it has low importance while growing the tree.

Post-pruning – once our tree is built to its depth, we can start pruning the nodes based on their significance.

Advantages:

- It produces a highly accurate classifier and learning is fast.
- It runs efficiently on large data bases.
- It can handle thousands of input variables without variable deletion.
- It computes proximities between pars of cases that can be used in clustering, locatingoutliers or give interesting views of the data.
- It offers an experimental method for detecting variable interactions.

Random Forest

Random Forest is a popular machine learning algorithm that belongs to the supervised learning technique. It can be used for both Classification and Regression problems in ML. It is based on the concept of ensemble learning, which is a process of combining multiple classifiers to solve a complex problem and to improve the performance of the model. As the name suggests, "Random Forest is a classifier that contains a number of decision trees on various subsets of the given dataset and takes the average to improve the predictive accuracy of that dataset." Instead of relying on one decision tree, the random forest takes the prediction from each tree and based on the majority votes of predictions, and it predicts the final output. The greater number of trees in the forest leads to higher accuracy and prevents the problem of overfitting.

The below diagram explains the working of the Random Forest algorithm:

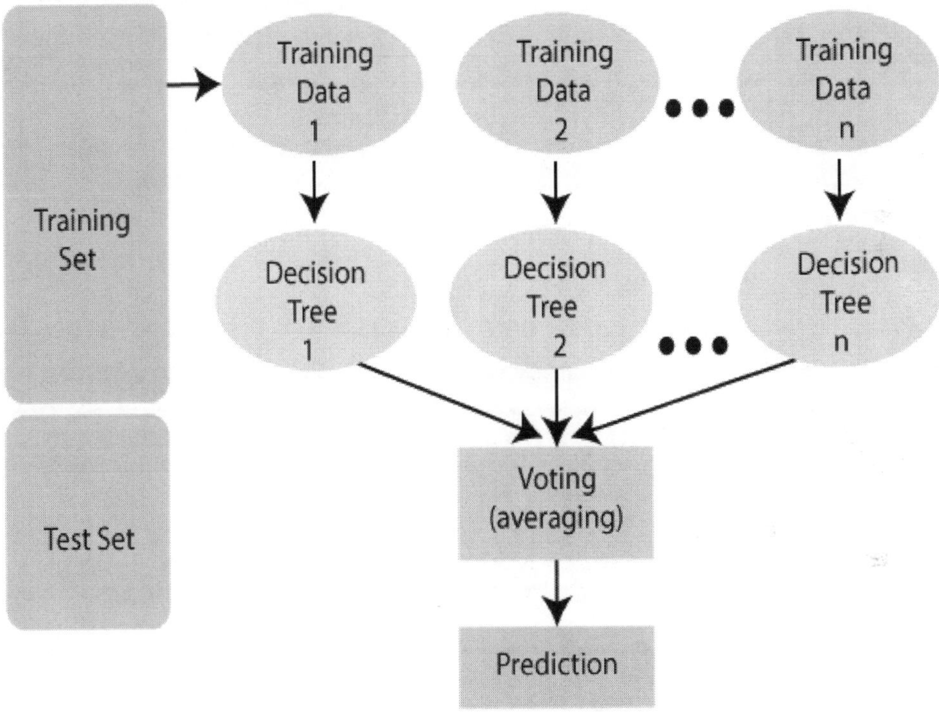

Figure 3.17: Working of the Random Forest algorithm

Why use Random Forest?

Below are some points that explain why we should use the Random Forest algorithm:

- It takes less training time as compared to other algorithms.
- It predicts output with high accuracy, even for the large dataset it runs efficiently.
- It can also maintain accuracy when a large proportion of data is missing.

How does Random Forest algorithm work?

Random Forest works in two-phase first is to create the random forest by combining Ndecision tree, and second is to make predictions for each tree created in the first phase.

The Working process can be explained in the below steps and diagram:Step-1: Select random K data points from the training set.

Step-2: Build the decision trees associated with the selected data points (Subsets).Step-3: Choose the number N for decision trees that you want to build.

Step-4: Repeat Step 1 & 2.

Step-5: For new data points, find the predictions of each decision tree, and assign the newdata points to the category that wins the majority votes.

The working of the algorithm can be better understood by the below example:

Example: Suppose there is a dataset that contains multiple fruit images. So, this dataset is given to the Random forest classifier. The dataset is divided into subsets and given to each decision tree. During the training phase, each decision tree produces a prediction result, and when a new data point occurs, then based on the majority of results, the Random Forest classifier predicts the final decision. Consider the below image:

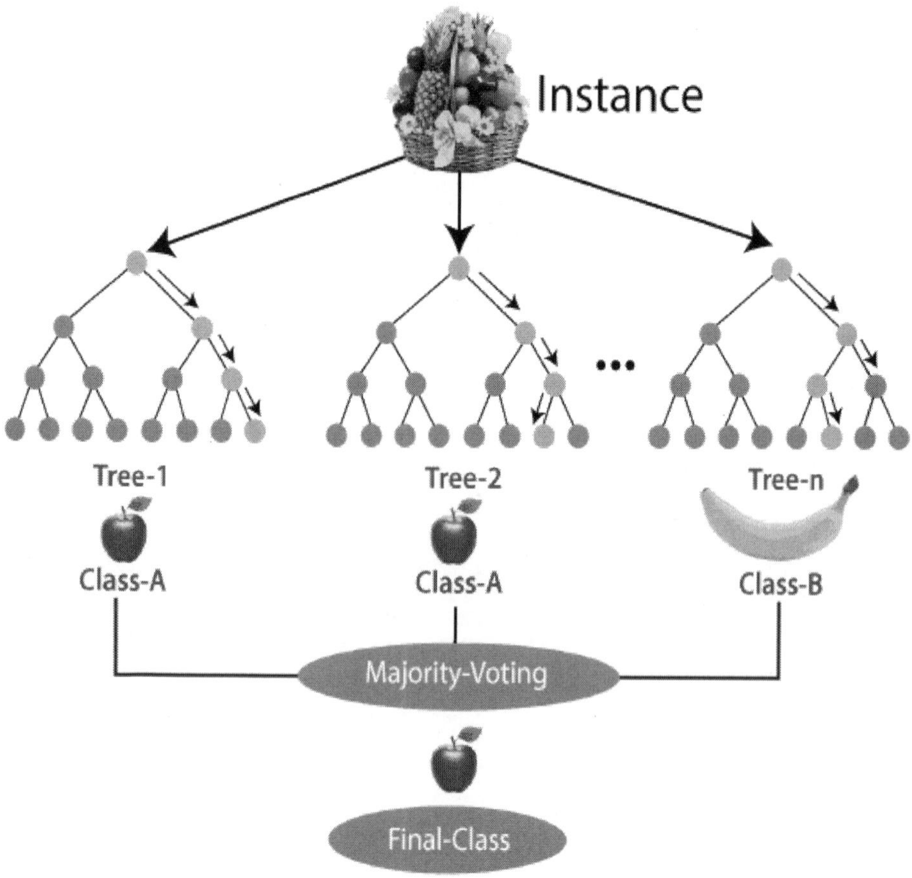

Figure 3.18: Example of working of the random forest

Applications of Random Forest:

There are mainly four sectors where Random forest mostly used:

- Banking: Banking sector mostly uses this algorithm for the identification of loan risk.
- Medicine: With the help of this algorithm, disease trends and risks of the disease can beidentified.
- Land Use: We can identify the areas of similar land use by this algorithm.
- Marketing: Marketing trends can be identified using this algorithm.
- Stock Market: Financial analysts use it to identify potential markets for stocks. It alsoenables them to remember the behaviour of stocks.
- E-Commerce: Through this system, e-commerce vendors can predict the preference of customers based on past consumption behaviour.

K Nearest Neighbour (KNN):

K-Nearest Neighbour is one of the simplest Machine Learning algorithms based on Supervised Learning technique. K-NN algorithm assumes the similarity between the new case/data and available cases and put the new case into the category that is most similar to the available categories. K-NN algorithm stores all the available data and classifies a new data point based on the similarity. This means when new data appears then it can be easily classified into a well suite category by using K- NN algorithm. K-NN algorithm can be used for Regression as well as for Classification but mostly it is used for the Classification problems. K-NN is a non-parametric algorithm, which means it does not make any assumption on underlying data. It is also called a lazy learner algorithm because it does not learn from the training set immediately instead it stores the dataset and at the time of classification, it performs an action on the dataset. KNN algorithm at the training

Figure 3.19: Identification of animal using KNN classifier

phase just stores the datasetand when it gets new data, then it classifies that data into a category that is much similar to the new data.

Example: Suppose, we have an image of a creature that looks similar to cat and dog, but we want to know either it is a cat or dog. So for this identification, we can use the KNN algorithm, as it works on a similarity measure. Our KNN model will find the similar features of the new data set to the cats and dogs images and based on the most similar features it will put it in either cat or dog category.

Why do we need a K-NN Algorithm?

Suppose there are two categories, i.e., Category A and Category B, and we have a new data point x1, so this data point will lie in which of these categories. To solve this type of problem, we need a K-NN algorithm. With the help of K-NN, we can easily identify the category or class of a particular dataset. Consider the below diagram:

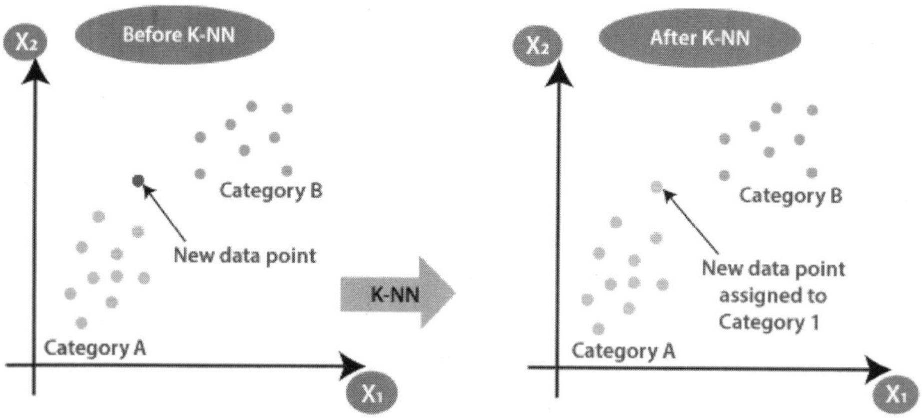

Figure 3.20: Classification of new data value using KNN

How does K-NN work?

The K-NN working can be explained on the basis of the below algorithm: Step-1: Select the number K of the neighbors

Step-2: Calculate the Euclidean distance of K number of neighbors

Step-3: Take the K nearest neighbors as per the calculated Euclidean distance.

Step-4: Among these k neighbors, count the number of the data points in each category.

Step-5: Assign the new data points to that category for which the number of the neighbor ismaximum.

Step-6: Our model is ready.

Suppose we have a new data point and we need to put it in the required category. Considerthe below image:

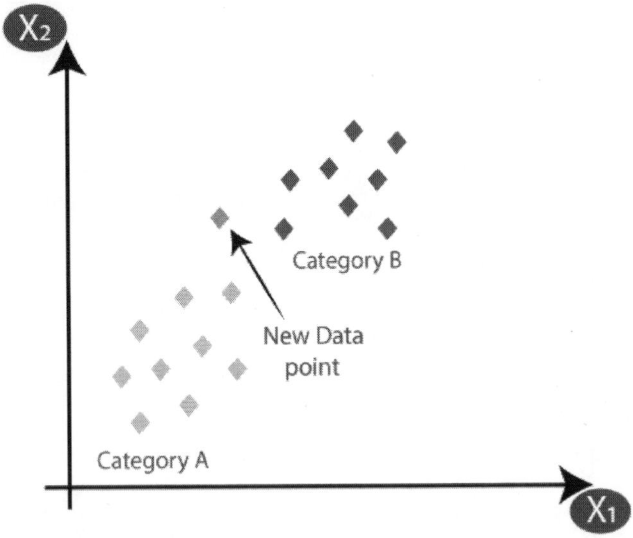

Firstly, we will choose the number of neighbors, so we will choose the k=5.

Next, we will calculate the Euclidean distance between the data points. The Euclidean distance is the distance between two points, which we have already studied in geometry. It can be calculated as:

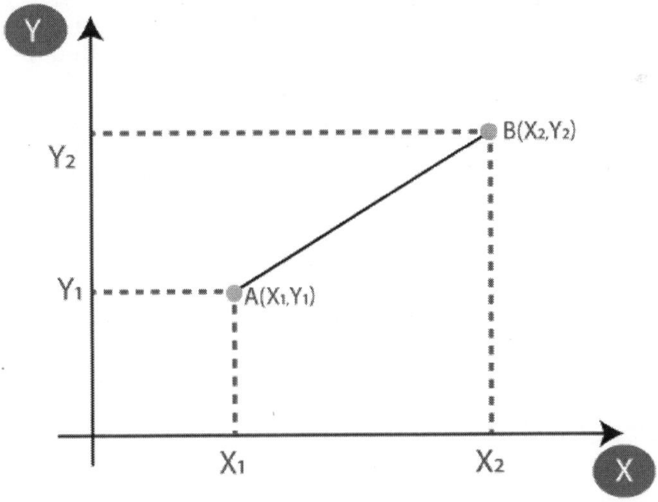

Euclidean Distance between A_1 and B_2 = $\sqrt{(X_2-X_1)^2+(Y_2-Y_1)^2}$

By calculating the Euclidean distance we got the nearest neighbors, as three nearest neighbors in category A and two nearest neighbors in category B. Consider the below image:

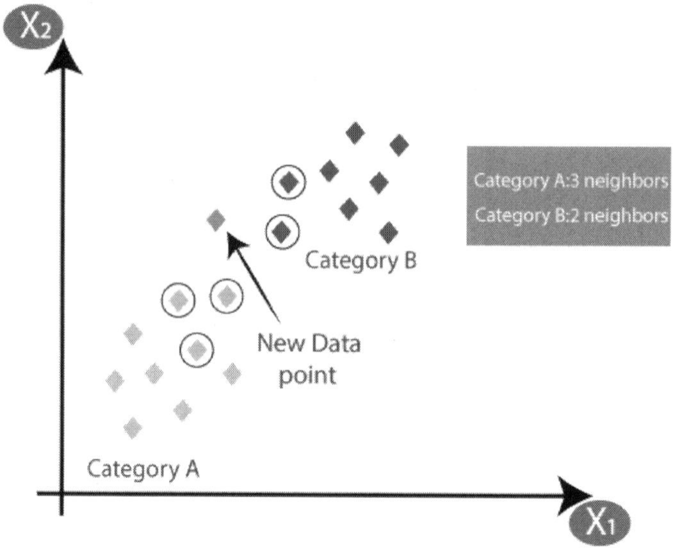

As we can see the 3 nearest neighbors are from category A, hence this new data pointmust belong to category A. How to select the value of K in the K-NN Algorithm?

- Below are some points to remember while selecting the value of K in the K-NNalgorithm:
- There is no particular way to determine the best value for "K", so we need to try some
- values to find the best out of them. The most preferred value for K is 5.
- A very low value for K such as K=1 or K=2, can be noisy and lead to the effects ofoutliers in the model.
- Large values for K are good, but it may find some difficulties.

Advantages of KNN Algorithm:

It is simple to implement.

It is robust to the noisy training data

It can be more effective if the training data is large.

K-Nearest Neighbors Classifiers and Model Example With Data Set:

Learn how to calculate the distance between a new entry and other existing values using theEuclidean distance formula.

Brightness	Saturation	Class
40	20	Red
50	50	Blue

60	90	Blue
10	25	Red
70	70	Blue
60	10	Red
25	80	Blue

The table above represents our data set. We have two columns —Brightness and Saturation.Each row in the table has a class of either Red or Blue.

Before we introduce a new data entry, let's assume the value of K is 5.

How to Calculate Euclidean Distance in the K-Nearest Neighbors Algorithm Here's the new data entry:

Brightness	Saturation	Class
20	35	?

We have a new entry but it doesn't have a class yet. To know its class, we have to calculate the distance from the new entry to other entries in the data set using the Euclidean distance formula.

Here's the formula: $(X_2\text{-}X_1)^2+(Y_2\text{-}Y_1)^2$ Where:

X_2 = New entry's brightness (20).

X_1= Existing entry's brightness.

Y_2 = New entry's saturation (35).

Y_1 = Existing entry's saturation.

Let's do the calculation together. I'll calculate the first three.Distance #1

For the first row, d1:

Brightness	Saturation	Class
20	35	?

d1 = √(20 - 40) + (35 - 20)

= √400 + 225

= √625

= 25

We now know the distance from the new data entry to the first entry in the table. Let's updatethe table.

Brightness	Saturation	Class	Distance
40	20	Red	25
50	50	Blue	?
60	90	Blue	?
10	25	Red	?
70	70	Blue	?
60	10	Red	?

Distance #2
For the second row, d2:

Brightness	Saturation	Class	Distance
50	50	Blue	?

d2 = √(20 - 50) + (35 - 50)

= √900 + 225

= √1125

= 33.54

Here's the table with the updated distance:

Brightness	Saturation	Class	Distance
40	20	Red	25
50	50	Blue	33.54
60	90	Blue	?
10	25	Red	?
70	70	Blue	?
60	10	Red	?
25	80	Blue	?

At this point, you should understand how the calculation works. Attempt to calculate thedistance for the last four rows.

Here's what the table will look like after all the distances have been calculated:

Brightness	Saturation	Class	Distance
40	20	Red	25
50	50	Blue	33.54
60	90	Blue	68.01

Brightness	Saturation	Class	Distance
10	25	Red	10
70	70	Blue	61.03
60	10	Red	47.17
25	80	Blue	45

Let's rearrange the distances in ascending order:

Brightness	Saturation	Class	Distance
10	25	Red	10
40	20	Red	25
50	50	Blue	33.54
25	80	Blue	45
60	10	Red	47.17
70	70	Blue	61.03
60	90	Blue	68.01

Since we chose 5 as the value of K, we'll only consider the first five rows. That is:

Brightness	Saturation	Class	Distance
10	25	Red	10
40	20	Red	25
50	50	Blue	33.54
25	80	Blue	45
60	10	Red	47.17

As you can see above, the majority class within the 5 nearest neighbors to the new entryis Red. Therefore, we'll classify the new entry as Red.

Here's the updated table:

Brightness	Saturation	Class
40	20	Red
50	50	Blue
60	90	Blue
10	25	Red
70	70	Blue

Brightness	Saturation	Class
60	10	Red
25	80	Blue
20	35	Red

What is Deep Learning Algorithm?

Deep learning can be defined as the method of machine learning and artificial intelligence that is intended to intimidate humans and their actions based on certain human brain functions to make effective decisions. It is a very important data science element that channels its modeling based on data-driven techniques under predictive modeling and statistics. To drive such a human-like ability to adapt and learn and to function accordingly, there have to be some strong forces which we popularly called algorithms.

Deep learning algorithms are dynamically made to run through several layers of neural networks, which are nothing but a set of decision-making networks that are pre-trained to serve a task. Later, each of these is passed through simple layered representations and move on to the next layer. However, most machine learning is trained to work fairly well on datasets that have to deal with hundreds of features or columns. For a data set to be structured or unstructured, machine learning tends to fail mostly because they fail to recognize a simple image having a dimension of 800x1000 in RGB. It becomes quite unfeasible for a traditional machine learning algorithm to handle such depths. This is where deep learning.

Importance of Deep Learning

Deep learning algorithms play a crucial role in determining the features and can handle the large number of processes for the data that might be structured or unstructured. Although, deep learning algorithms can overkill some tasks that might involve complex problems because they need access to huge amounts of data so that they can function effectively. For example, there's a popular deep learning tool that recognizes images namely Imagenet that has access to 14 million images in its dataset-driven algorithms. It is a highly comprehensive tool that has defined a next-level benchmark for deep learning tools that aim images as their dataset.

Deep learning algorithms are highly progressive algorithms that learn about the image that we discussed previously by passing it through each neural network layer. The layers are highly sensitive to detect low-level features of the image like edges and pixels and henceforth the combined layers take this information and form holistic representations by comparing it with previous data. For example, the middle layer might be programmed to detect some special parts of the object in the photograph which other deep trained layers are programmed to detect special objects like dogs, trees, utensils, etc.

However, if we talk out the simple task that involves less complexity and a data-driven resource, deep learning algorithms fail to generalize simple data. This is one of the main reasons deep learning is not considered effective as linear or boosted tree models. Simple models aim to churn out custom data, track fraudulent transactions and deal with less complex datasets with fewer features. Also, there are various cases like multiclass classification where deep learning can be effective because it involves smaller but more structured datasets but is not preferred usually.

Deep Learning is a part of Machine Learning used to solve complex problems and build intelligent solutions. The core concept of Deep Learning has been derived from the structure and function of the human brain. Deep Learning uses artificial neural networks to analyze data and make predictions. It has found its application in almost every sector of business. In this article, we'll learn the top Deep Learning

Deep Learning Applications:

1. Virtual Assistants

Virtual Assistants are cloud-based applications that understand natural language voice commands and complete tasks for the user. Amazon Alexa, Cortana, Siri, and Google Assistant are typical examples of virtual assistants. They need internet-connected devices to work with their full capabilities. Each time a command is fed to the assistant, they tend to provide a better user experience based on past experiences using Deep Learning algorithms.

2. Chatbots

Chatbots can solve customer problems in seconds. A chatbot is an AI application to chat online via text or text-to-speech. It is capable of communicating and performing actions similar to a human. Chatbots are used a lot in customer interaction, marketing on social network sites, instant messaging the client. It delivers automated responses to user inputs. It uses machine learning and deep learning algorithms to generate different types of reactions.

3. Healthcare

Deep Learning has found its application in the Healthcare sector. Computer-aided disease detection and computer-aided diagnosis have been possible using Deep Learning. It is widely used for medical research, drug discovery, and diagnosis of life-threatening diseases such as cancer and diabetic retinopathy through the process of medical imaging.

4. Entertainment

Companies such as Netflix, Amazon, YouTube, and Spotify give relevant movies, songs, andvideo recommendations to enhance their customer experience. This is all thanks to Deep Learning. Based on a person's browsing history, interest, and behavior, online streaming companies give suggestions to help them make product and service choices. Deep learning techniques are also used to add sound to silent movies and generate subtitles automatically.

5. Robotics

Deep Learning is heavily used for building robots to perform human-like tasks. Robots powered by Deep Learning use real-time updates to sense obstacles in their path and pre- plan their journey instantly. It can be used to carry goods in hospitals, factories, warehouses, inventory management, manufacturing products, etc.

6. Natural Language Processing

Another important field where Deep Learning is showing promising results is NLP, It is the procedure for allowing robots to study and comprehend human language. Machines are discouraged from correctly comprehending or creating human language not only because of the alphabet and words, but also because of context, accents, handwriting, and other factors.

Artificial Neural Network (ANN)

Introduction to Artificial Neural Networks:

Artificial Neural Networks are the most popular machine learning algorithms today. The invention of these Neural Networks took place in the 1970s but they have achieved huge popularity due to the recent increase in computation power because of which they are now virtually everywhere. In every application that you use, Neural Networks power the intelligent interface that keeps you engaged.

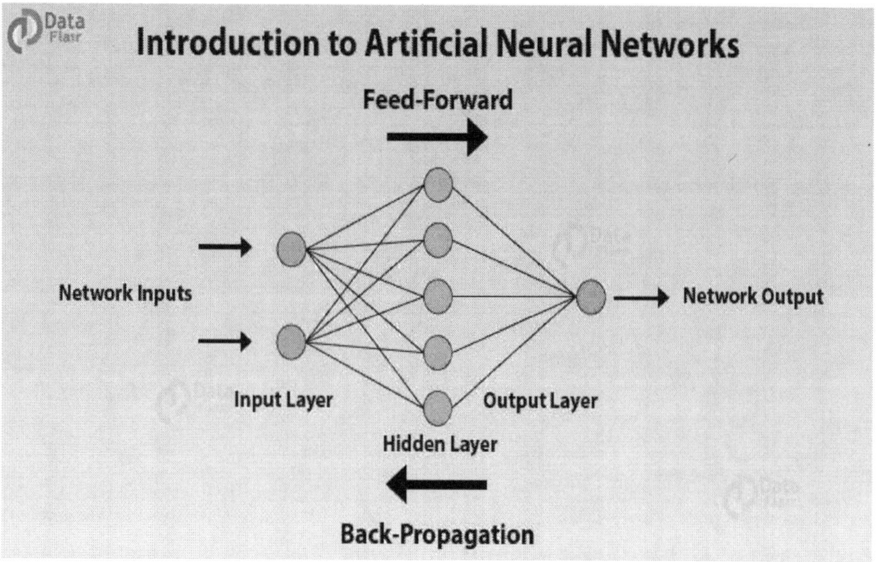

Figure 3.21: Artifical neural network

What is ANN?

Artificial Neural Networks are a special type of machine learning algorithms that are modeled after the human brain. That is, just like how the neurons in our nervous system are able to learn from the past data, similarly, the ANN is able to learn from the data and provide responses in the form of predictions or classifications.

ANNs are nonlinear statistical models which display a complex relationship between the inputs and outputs to discover a new pattern. A variety of tasks such as image recognition, speech recognition, machine translation as well as medical diagnosis makes use of these artificial neural networks.

An important advantage of ANN is the fact that it learns from the example data sets. Most commonly usage of ANN is that of a random function approximation. With these types of tools, one can have a cost-effective method of arriving at the solutions that define the distribution. ANN is also capable of taking sample data rather than the entire dataset to provide the output result. With ANNs, one can enhance existing data analysis techniques owing to their advanced predictive capabilities.

Artificial Neural Networks Architecture

The functioning of the Artificial Neural Networks is similar to the way neurons work in our nervous system. The Neural Networks go back to the early 1970s when Warren S McCulloch and Walter Pitts coined this term. In order to understand the workings of ANNs, let us first understand how it is structured. In a neural network, there are three essential layers –

Input Layers:

The input layer is the first layer of an ANN that receives the input information in the form of various texts, numbers, audio files, image pixels, etc.

Hidden Layers:

In the middle of the ANN model are the hidden layers. There can be a single hidden layer, as in the case of a perceptron or multiple hidden layers. These hidden layers perform various types of mathematical computation on the input data and recognize the patterns that are part.

Output Layer:

In the output layer, we obtain the result that we obtain through rigorous computationsperformed by the middle layer.

In a neural network, there are multiple parameters and hyperparameters that affect the performance of the model. The output of ANNs is mostly dependent on these parameters. Some of these parameters are weights, biases, learning rate, batch size etc. Each node in the ANN has some weight.

Each node in the network has some weights assigned to it. A transfer function is used for calculating the weighted sum of the inputs and the bias.

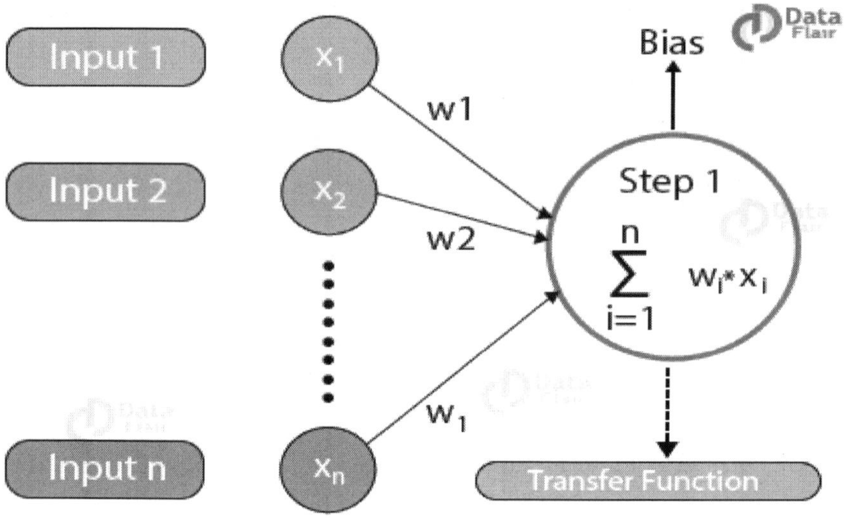

Figure 3.22: Step 1 - Calculation of weighted sum of input and the bias using transfer function

After the transfer function has calculated the sum, the activation function obtains the result. Based on the output received, the activation functions fire the appropriate result from the node. For example, if the output received is above 0.5, the activation function fires a 1 otherwise it remains 0.

Some of the popular activation functions used in Artificial Neural Networks are Sigmoid, RELU, Softmax, tanh etc.

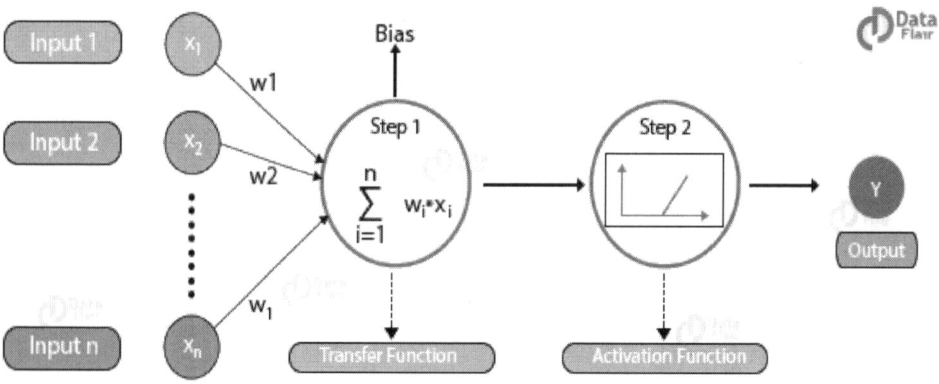

Figure 3.23: Step 2 – Activation function and the output

Based on the value that the node has fired, we obtain the final output. Then, using the error functions, we calculate the discrepancies between the predicted output and resulting output and adjust the weights of the neural network through a process known as backpropagation.

ANNs are part of an emerging area in Machine Learning known as Deep Learning.

Many people are confused between Deep Learning and Machine Learning. Are you among one of them? Check this easy to understand article on Deep Learning vs Machine Learning.

Back Propagation in Artificial Neural Networks:

In order to train a neural network, we provide it with examples of input-output mappings. Finally, when the neural network completes the training, we test the neural network where we do not provide it with these mappings. The neural network predicts the output and we evaluate how correct the output is using the various error functions. Finally, based on the result, the model adjusts the weights of the neural networks to optimize the network following gradient descent through the chain rule.

Types of Artificial Neural Networks

There are two important types of Artificial Neural Networks –

- FeedForward Neural Network
- FeedBack Neural Network
- FeedForward Artificial Neural Networks

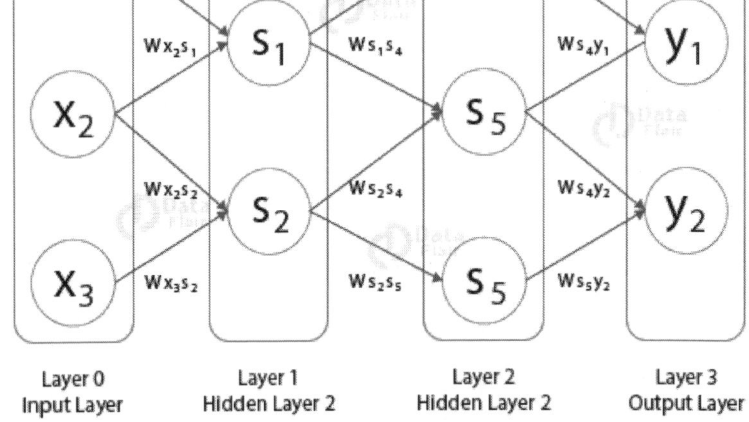

In the feedforward ANNs, the flow of information takes place only in one direction. That is, the flow of information is from the input layer to the hidden layer and finally to the output. There are no feedback loops present in this neural network. These type of neural networks are mostly used in supervised learning for instances such as classification, image recognition etc. We use them in cases where the data is not sequential in nature.

Feedback Artificial Neural Networks:

In the feedback ANNs, the feedback loops are a part of it. Such type of neural networks are mainly for memory retention such as in the case of recurrent neural networks. These types ofnetworks are most suited for areas where the data is sequential or time-dependent.

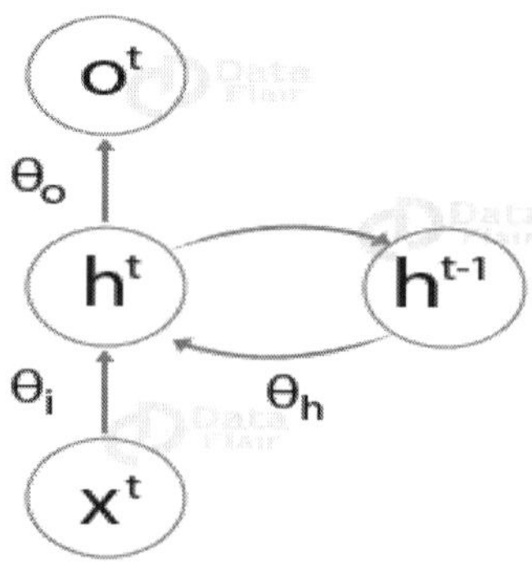

Artificial Neural Networks Applications:

Following are the important Artificial Neural Networks applications – Handwritten Character Recognition:

ANNs are used for handwritten character recognition. Neural Networks are trained torecognize the handwritten characters which can be in the form of letters or digits.

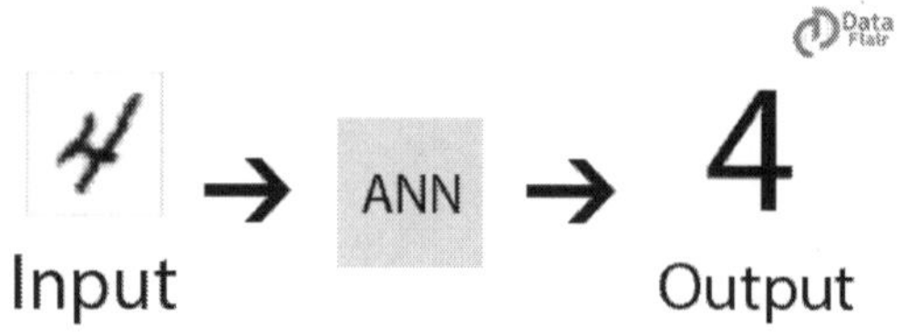

Speech Recognition:

ANNs play an important role in speech recognition. The earlier models of Speech Recognition were based on statistical models like Hidden Markov

Models. With the advent of deep learning, various types of neural networks are the absolute choice for obtaining an accurate classification.

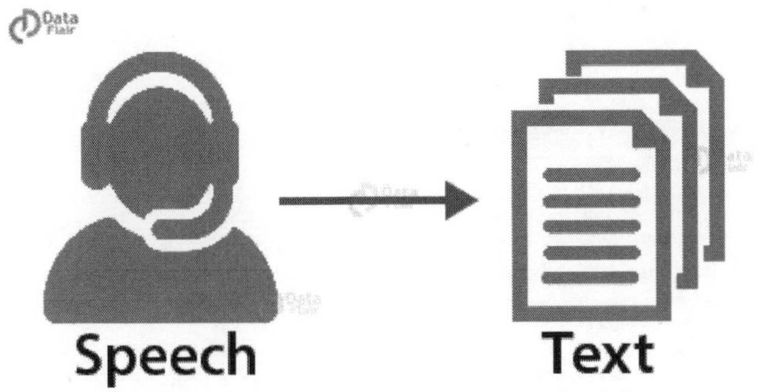

Advantages:

- Parallel processing capability
- Storing data on the entire network
- Capability to work with incomplete knowledge
- Having a memory distribution
- Having fault tolerance

Convolutional Neural Network (CNN):

Convolutional Neural Network is one of the technique to do image classification and image recognition in neural networks. It is designed to process the data by multiple layers of arrays.This type of neural network is used in applications like image recognition or face recognition. The primary difference between CNN and other neural network is that CNN takes input as a two-dimensional array. And it operates directly on the images rather than focusing on feature extraction which other neural networks do.

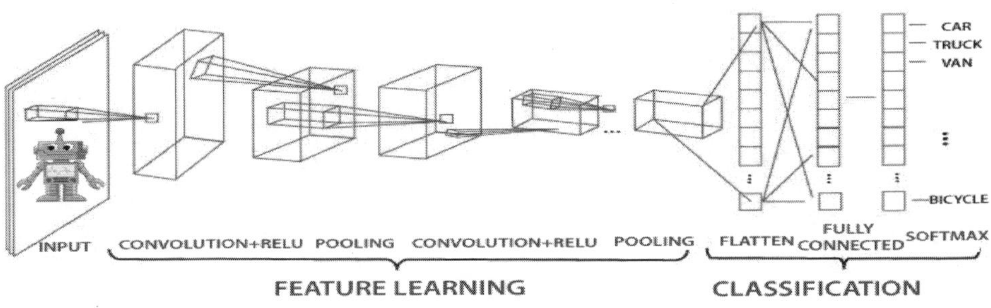

The dominant approach of CNN includes the solution for problems of recognition. Some companies like Google and Facebook have invested in the field research and development concerning recognition projects to get activities done with higher speed.

Scene labeling, object detection, and face recognition, etc. are some of the areas where

Convolutional Neural Network works.

Convolutional Neural Network (CNN or ConvNet) is a type of feed-forward artificial network where the connectivity pattern between its neurons is inspired by the organization ofthe animal visual cortex.

Working of CNN:

Generally, A Convolutional neural network has three layers. And we understand each layer one by one with the help of an example of the classifier. With it can classify an image of an X and O. So, with the case, we will understand all four layers.

Convolutional Neural Networks have the following layers:

- Convolutional
- ReLU Layer
- Pooling
- Fully connected layer

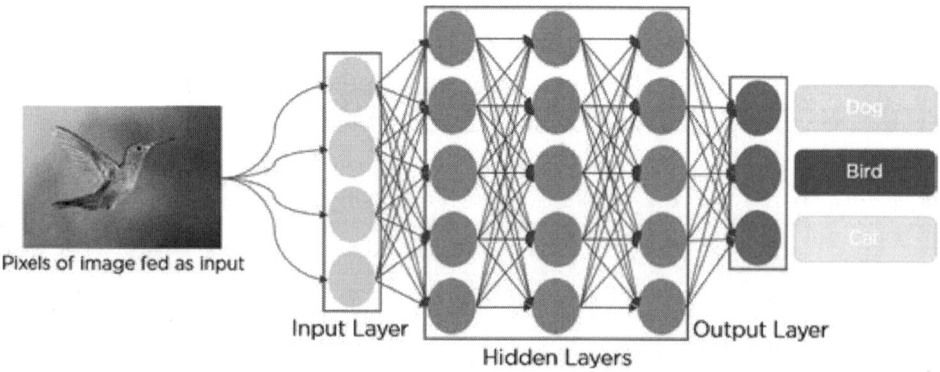

Convolution Layer:

This is the first step in the process of extracting valuable features from an image. A convolution layer has several filters that perform the convolution operation. Every image is considered as a matrix of pixel values.

ReLU layer:

ReLU stands for the rectified linear unit. Once the feature maps are extracted, the next step isto move them to a ReLU layer.

Pooling Layer:

Pooling is a down-sampling operation that reduces the dimensionality of the feature map.The rectified feature map now goes through a pooling layer to generate a pooled feature map.

Fully connected layer:

The next step in the process is called flattening. Flattening is used to convert all the resultant 2-Dimensional arrays from pooled feature maps into a single long continuous linear vector.

Here's how exactly CNN recognizes a bird:

The pixels from the image are fed to the convolutional layer that performs the convolutionoperation.

It results in a convolved map.

The convolved map is applied to a ReLU function to generate a rectified feature map.

The image is processed with multiple convolutions and ReLU layers for locating thefeatures.

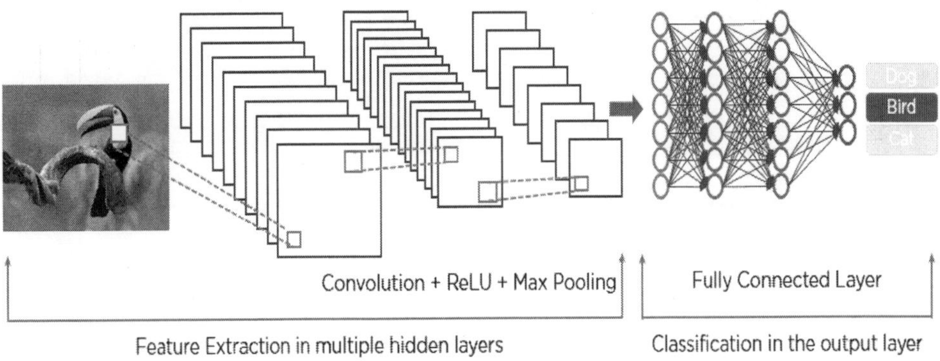

Different pooling layers with various filters are used to identify specific parts of the image. The pooled feature map is flattened and fed to a fully connected layer to get the final output.

References

Xiao, T., Xia, T., Yang, Y., Huang, C. and Wang, X., 2015. Learning from massive noisy labeled data for image classification. *In Proceedings of the IEEE conference on computer vision and pattern recognition (pp. 2691-2699).*

Ying, X., 2019, February. An overview of overfitting and its solutions. In *Journal of physics: Conference series* (Vol. 1168, p. 022022). IOP Publishing.

Sen, P.C., Hajra, M. and Ghosh, M., 2020. Supervised classification algorithms in machine learning: A survey and review. *In Emerging Technology in Modelling and Graphics: Proceedings of IEM Graph 2018* (pp. 99-111). Springer Singapore.

Karaca, K., 2021. Values and inductive risk in machine learning modelling: the case of binary classification models. *European Journal for Philosophy of Science, 11*(4), p.102.

Nasteski, V., 2017. An overview of the supervised machine learning methods. *Horizons. b*, 4(51-62), p.56.

CHAPTER 4

SPATIO-TEMPORAL MONITORING OF WATER SURFACE TEMPERATURE USING GOOGLE EARTH ENGINE

Mohammed Meharoof[1,2], Vinod Kumar Yadav[2], and Chandani Dave[2]

1 Assistant Professor (o/c), College of Fisheries, Payyanur, Kannur-670307
2 ICAR-Central Institute of fisheries Education, Mumbai-400061

Geographic Information Systems and Remote Sensing

Geographic Information Systems (GIS) and Remote Sensing are two powerful technologies that have revolutionized the way we perceive, analyze, and manage our world. These interrelated fields have rapidly evolved, becoming essential tools across a multitude of disciplines, from environmental science and urban planning to disaster management and agriculture. At their core, GIS and Remote Sensing are instrumental in harnessing spatial data to gain a deeper understanding of our planet and to make informed decisions that impact our environment and society. GIS, often referred to as the "science of where," involves the collection, storage, analysis, and visualization of geographic information. It allows us to represent the real world in a digital format, where various layers of spatial data, such as maps, satellite imagery, and socioeconomic data, are integrated and manipulated to answer complex questions. GIS provides the means to explore patterns, relationships, and trends within this data, offering valuable insights into both the natural and human-made aspects of our planet. Remote Sensing, on the otherhand, is the science of obtaining information about the Earth's surface without physical contact. This is achieved by using sensors mounted on aircraft or satellites to capture data in the form of images or other measurements. These remote sensing platforms provide a unique perspective, enabling us to observe large areas from above and collectdata in wavelengths beyond what the human eye can perceive. Remote sensing technology has made it possible to monitor changes in land cover, detect environmental anomalies, and track dynamic processes, such as deforestation, urban sprawl, and climate patterns. Geographic Information Systems (GIS) and Remote Sensing represent two pivotal disciplines at the intersection of spatial science and technology. These fields have profoundly influenced diverse domains, ranging from environmental management and urban planning to disaster response and agriculture. This introduction seeks to provide an overview of GIS and Remote Sensing, their significance, and their widespread applications, with reference to seminal works and established authorities in the field.

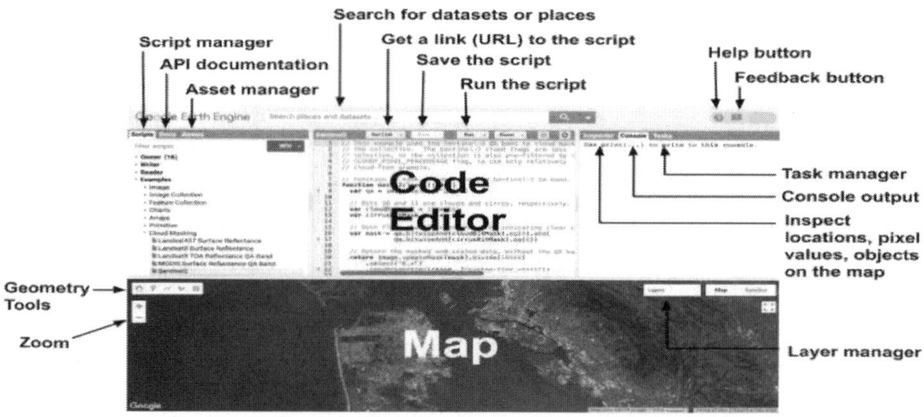

Figure 4.1: Google Earth Engine Workspace

Geographic Information Systems (GIS) is a multidisciplinary frameworkfor capturing, storing, analyzing, and visualizing geospatial data (Longley et al., 2015). It revolves around the core concept of location, facilitating the integration of various data layers, such as maps, satellite images, and socioeconomic statistics, to gain insights into the spatial relationships and patterns within our world (Burrough and McDonnell, 2015). This technology has become indispensable for decision-making processes across sectors (Tomlin, 2013). Notably, the book "Geographic Information Science and Systems" by Longley et al.

(2015) offers a comprehensive introduction to GIS principles and applications. Remote Sensing is the science of acquiring information about Earth's surface from a distance, typically through sensors aboard satellites or aircraft (Lillesand et al.,

2015). It provides a unique perspective, allowing the collection of data across a broad spectrum of wavelengths, including those beyond the visible spectrum (Jensen, 2009). Remote sensing data enables the monitoring of land cover changes, environmental phenomena, and dynamic processes on a global scale (Lillesand et al., 2015). Jensen's book, "Remote Sensing of the Environment: AnEarth Resource Perspective" (2009), is a foundational reference for understanding the principles and applications of remote sensing. The synergy between GIS and Remote Sensing is a defining feature of contemporary spatial science (Campbell, 2011). Remote sensing platforms provide the raw data, which GIS then incorporates to create actionable insights (Voigt and Yang, 2015). As the influence of GIS and Remote Sensing continues to expand, they remain essential tools for addressing pressing global challenges such as climate change, resource management, and sustainable development (Pettorelli et al., 2014; Turner etal., 2015). This introduction serves as a gateway to the extensive and evolving domains of GIS and Remote Sensing, guided by foundational works and expert insights.

Google Earth Engine (GEE)

Google Earth Engine is one of the most transformative tools in the realm of geospatial analysis and remote sensing. Developed by Google, this cloud-based platform has revolutionized the way we access, process, and analyze petabytes of Earth observation data for a wide range of applications, from environmental monitoring to disaster management andbeyond. Over the past few decades, Earth observation satellites have continuously captured an unprecedented volume of data, providing a wealth of information about our planet's dynamic processes, ecosystems, and changes over time. This influx of data presented both anopportunity and a challenge—how to efficiently manage, process, and extract meaningful insights from this immense pool of information. Google Earth Engine emerged as a powerfulsolution to this challenge. At its core, Google Earth Engine is a geospatial data processing and analysis platform that hosts an extensive archive of satellite and Earth observation imagery. It seamlessly integrates data from various sources, including Landsat, Sentinel, MODIS, and more, into a unified and easily accessible interface. This rich collection of imagery, spanning decades, allows researchers, scientists, and analysts to investigate changesin land cover, monitor deforestation, track urban expansion, assess agricultural trends, and explore numerous other Earth-related phenomena. What sets Google Earth Engine apart is itscapacity to perform geospatial analysis at an unprecedented scale. With its powerful computing infrastructure, it enables users to apply complex algorithms and machine learning models to large and continuous datasets. This capability opens a world of possibilities for tackling global challenges, such as climate change, food security, and disaster risk reduction. Google Earth Engine follows a cloud-based model, eliminating the need for users to invest inhigh-end computing resources or manage large datasets locally. This accessibility fosters collaboration and innovation within the geospatial community, enabling researchers and organizations worldwide to work together on shared goals. It also provides a valuable resource for educational institutions, allowing students to gain hands-on experience in geospatial analysis. The versatility of Google Earth Engine is evident in its applications across diverse fields. Ecologists use it to study habitat changes, climatologists track weather patterns, epidemiologists monitor disease outbreaks, and urban planners assess urban growth. The platform has been instrumental in disaster response, aiding in damage assessment and recovery efforts during wildfires, hurricanes, and earthquakes. Google Earth Engine represents a paradigm shift in how we harness the power of Earth observation data. Its user- friendly interface, massive data archive, and scalable computing capabilities make it an indispensable tool for both seasoned geospatial professionals and newcomers to the field. As a GIS expert and professor, I encourage you to explore the immense potential of Google Earth Engine and witness how it continues to shape the future of geospatial science and our understanding of the Earth's dynamic processes.

Advantages of GEE

1. Massive Geospatial Data Archive: GEE provides access to an extensive and regularly updated archive of Earth observation data from various satellite missions. This includes data on land cover, climate, vegetation, urban development, and more. Having this vast repository at your fingertips saves time and effort in data acquisition
2. Cloud-Based Computing: GEE harnesses the computational power of Google's cloud infrastructure, allowing users to analyze large-scale geospatial datasets without the need for high-end hardware. This means you can perform complex analyses, such as time-series analysis and machine learning, quickly and efficiently
3. Global Coverage: GEE provides global coverage, meaning you can analyze data from anywhere on Earth. This is especially valuable for researchers interested in studying phenomena at regional, national, or global scales
4. Integration of Remote Sensing and GIS: GEE seamlessly integrates remote sensing and geographic information system (GIS) capabilities. You can perform both spatial and temporal analyses, making it a versatile tool for monitoring changes over time
5. Time-Series Analysis: GEE excels in time-series analysis, allowing you to track changes in geospatial data over extended periods. This is particularly valuable for studying land use change, deforestation, urban growth, and climate trends
6. Machine Learning and AI: GEE provides built-in support for machine learning and artificial intelligence (AI) techniques. You can train models using Earth observation data to perform tasks like land cover classification, object detection, and prediction
7. Visualization Tools: GEE offers interactive visualization tools that make it easy to create maps, charts, and graphs from your analysis results. These visualizations are valuable for communicating your findings effectively

8. Open Source and Collaborative: GEE supports open-source scripting using JavaScript and Python, making it accessible to a wide range of users. It also encourages collaboration by allowing users to share their scripts and analyses with others
9. Environmental Monitoring and Conservation: GEE has been used extensively in environmental monitoring and conservation efforts. It aids in tracking deforestation, monitoring wildlife habitats, assessing land degradation, and responding to natural disasters
10. Policy and Decision Support: GEE's data and analysis capabilities support informed decision-making in areas such as agriculture, forestry, urban planning, and disaster management
11. Cost-Efficient: Access to GEE is often free for non-commercial, research, and educational purposes. This significantly reduces the cost barriers associated with geospatial analysis

Steps to register on google earth engine

Step 1: Access the Google Earth Engine Website

Open your web browser and go to the Google Earth Engine website:https://earthengine.google.com/

Step 2: Sign In with Your Google Account

Click on the "Sign In" button located in the upper right corner of the page. Step 3: Review and Agree to the Terms of Service

Before you can proceed, you will be asked to review and accept the Terms of Service.Carefully read through the terms, and if you agree, click on the "Accept" button.

Step 4: Request Access

After accepting the Terms of Service, you will be taken to the Earth Engine dashboard. Here, you will find information about your Earth Engine account status. If you see a "Request Access" button, click on it.

Step 5: Fill in the Access Request Form

You will be directed to a form where you need to provide information about your intended use of Google Earth Engine. This form will ask you about your affiliation, research goals, and how you plan to use Earth Engine. Be specific and clear in your responses, as this will help Google assess your request.

Step 6: Submit Your Access Request

After completing the form, click the "Submit" or "Request Access" button to send your request to Google for review.

Step 7: Wait for Approval

Google will review your access request. The approval process can take some time, and it depends on the nature of your request and Google's current workload. Be patient during this stage.

Step 8: Receive Approval Email

Once your request is approved, you will receive an email notification from Google Earth Engine. This email will contain instructions on how to access and use Earth Engine.

Step 9: Access Earth Engine

Return to the Google Earth Engine website and sign in with your Google account. You should now have access to Earth Engine's powerful geospatial analysis tools and datasets.

Assessment of Water Surface Temperature using GEE

Surface waters, including rivers, lakes, oceans, and wetlands, play a pivotal role in maintaining the delicate ecological balance and hydrological cycle of our environment (Zhang et al., 2016). However, the dual forces of climate change and human activities pose significant threats to the stability of surface waters, leading to both interannual and intra- annual variations (Xia et al., 2019). These fluctuations can have far-reaching consequences for both human societies and ecosystems, making it imperative to closely monitor and understand these changes. One critical aspect of this monitoring process involves mapping the spatio-temporal extent of water surfaces. Such assessments are invaluable for conducting ecological and environmental impact analyses. Moreover, having access to reliable data on the dynamics of surface waters empowers decision-makers to implement effective measures for their protection and fosters the development of sustainable long-term management strategies (Albarqouni et al., 2022). Remote sensing (RS) emerges as a cost-effective and highly efficient tool for the comprehensive monitoring of spatial and temporal changes in surface water resources. What sets RS apart is its capability to collect data without the need for direct contact with the water bodies or the ground beneath them (Asokan & Anitha, 2019). By harnessing the power of satellites and specialized sensors, RS enables us to observe and analyze variations in water surface extent over time, providing crucial insights into environmental dynamics. In addition to monitoring changes in surface water extent, assessing water surface temperature holds immense importance. Temperature is a fundamental parameter affecting the physical and biological properties of aquatic ecosystems. It influences the distribution and behavior of aquatic species, impacts water chemistry, and plays a role in the formation and dynamics of weather patterns. Monitoring water surface temperature through remote sensing helps in understanding and predicting these complex interactions. Furthermore, changes in water temperature can serve as early indicators of environmental disturbances, making it an essential component of comprehensive water resource management and environmental monitoring efforts.

Importance of WST Assessment

- Climate Change Monitoring: WST data provides essential information for monitoring and understanding the impacts of climate change. Rising temperaturesin oceans, lakes, and rivers are key indicators of global warming, which can have profound effects on ecosystems, weather patterns, and sea levels.
- Aquatic Ecosystem Health: WST affects the health and vitality of aquatic ecosystems. Many aquatic species, including fish and invertebrates, have specific temperature requirements for survival and reproduction. Deviations from these temperature ranges can lead to changes in species composition and ecosystem structure.
- Water Quality Assessment: WST can serve as a proxy for assessing water quality.
- Elevated temperatures can indicate pollution or nutrient runoff, which can have detrimental effects on aquatic life and human health. Monitoring WST helpsidentify areas where corrective actions are needed.
- Habitat Management: WST data is crucial for habitat management and conservation efforts. It helps identify critical habitats for species vulnerable to temperature changes and informs decisions related to protected areas, restorationprojects, and conservation priorities.
- Hydrological Modelling: Accurate WST data is essential for hydrological modeling.It influences the rate of evaporation, water density, and water movement in rivers and lakes, which, in turn, affects the availability of freshwater resources for human consumption and agriculture.
- Weather Forecasting: WST is a key parameter for weather forecasting and the prediction of extreme weather events. It influences local weather patterns, including the formation and intensity of storms, and can affect precipitation patterns.
- Public Health: WST data is relevant to public health. High water temperatures can create conditions conducive to the growth of harmful microorganisms in recreational waters, increasing the risk of waterborne diseases. It's also relevant formanaging the impacts of heatwaves on human populations.
- Agriculture: WST information is valuable for agriculture, especially for irrigation management. It helps farmers optimize irrigation schedules

based on temperature and evaporation rates, leading to more efficient water use.

- Fisheries Management: Fisheries managers use WST data to regulate fishing seasons and quotas. Temperature influences fish behavior, migration patterns, andspawning, making it a critical factor in sustainable fisheries management.
- Infrastructure Planning: Engineers and urban planners use WST data to design and maintain infrastructure related to water supply, wastewater treatment, and flood control. It helps ensure the resilience and functionality of these systemsunder changing temperature conditions.
- Research and Education: WST data supports scientific research in various fields, fromlimnology and oceanography to ecology and environmental science. It also plays
- a role in educating students and the public about the importance of water temperaturein environmental processes.

Requirements to Analyse WST in GEE

1. **Shape file of the Water boundary layer:** The outer boundary of the waterbody is made by visual interpretation by discriminating the farthest water mark of a waterbody visible as wet areas from the bordering dry area. This can be prepared fromQGIS/ArcGIS/Google Earth.
2. **Dataset:** The Landsat TM/ETM+/OLI satellite remote sensing images of the United

Bands	Wavelength (micrometers)	Resolution (meters)
Band 1 - Coastal aerosol	0.43-0.45	30
Band 2 - Blue	0.45-0.51	30
Band 3 - Green	0.53-0.59	30
Band 4 - Red	0.64-0.67	30
Band 5 - Near Infrared (NIR)	0.85-0.88	30
Band 6 - Shortwave Infrared (SWIR) 1	1.57-1.65	30
Band 7 - Shortwave Infrared (SWIR) 2	2.11-2.29	30
Band 8 - Panchromatic	0.50-0.68	15
Band 9 - Cirrus	1.36-1.38	30
Band 10 - Thermal Infrared (TIRS) 1	10.6-11.19	100
Band 11 - Thermal Infrared (TIRS) 2	11.50-12.51	100

Fig.4.2. Spectral Bands of Landsat 8 Operational Land Imager (OLI) and Thermal InfraredSensor (TIRS)

States Geological Survey (USGS) has been for this. The sentinel 2 multispectral imageries by European Space Agency can also be used for the study. Landsat 8 Operational Land Imager (OLI) and Thermal Infrared Sensor (TIRS) images consist of nine spectral bands with a spatial resolution of 30 meters for Bands 1 to 7 and 9.

3. **Water body extraction:** The Modified Normalized Difference Water Index (MNDWI) method is used to extract and separate the water and non-water pixels (Ji et al., 2022) as explained below. Modified Normalized Difference Water Index (MNDWI) effectively differentiates between water and urban areas in satellite images. This method utilizes the visible green (GREEN) and short-wave infrared 1 (SWIR1) spectral bands (Xu, 2006). They are the digital number value of green and Short Wave Infra-red 1 (SWIR1) bands of the satellite imagery. Range of MNDWI is between -1 to +1, where water features are indicated by positive values while the vegetation and soil features have the value between – 1 and 0.

4. **Top of Atmospheric (TOA) Radiance:** It refers to the radiance, or radiant energy, measured by sensors or instruments at the outer boundary of Earth's atmosphere. It represents the amount of electromagnetic radiation (light and heat) coming from the Earth's surface or other objects on or near the Earth's surface as it passes through the atmosphere and reaches the sensor.

   Where,

   M_L=Multiplicative rescaling factor of Band 10Q_{cal}=Band 10

   A_L=Additive rescaling factor of Band 10

5. **Brightness Temperature:** It refers to the temperature of a hypothetical blackbody (perfect emitter and absorber of radiation) that would emit the same amount of radiation as a given object, at a specific wavelength or frequency, observed froma remote sensor

   Brightness temperature (BT)=$(K_2/(\ln(K_1/TOA)+1))-273.15$

   Where,

   K_1=Thermal conversion constant one of band 10

   K_2=Thermal conversion constant two of band 10

6. **ractional vegetation:** It refers to the proportion or fraction of surface covered by vegetation

$$Fractional\ of\ vegetation\ (FV) = \left(\frac{(MNDWI - MNDWI_{min})}{(MNDWI_{max} - MNDWI_{min})}\right)^2$$

7. **Emissivity:** It is the radiation capacity of a surface compared to that of a black body Emmissivity ( )=0.004*FV+0.986

$$Water\ Surface\ Temperatue\ (WST) = \left(\frac{BT}{\left(1 + \left(0.00115 * \frac{BT}{1.4388}\right) * Ln(\varepsilon)\right)} \right)$$

8. Water Surface Temperature (WST):

References

1. Albarqouni, M. M., Yagmur, N., Bektas Balcik, F., & Sekertekin, A. (2022). Assessment of Spatio-Temporal Changes in Water Surface Extents and Lake Surface Temperatures Using Google Earth Engine for Lakes Region, Türkiye. ISPRS International Journal of Geo-Information, 11(7), 407.
2. Asokan, A., & Anitha, J. J. E. S. I. (2019). Change detection techniques for remote sensing applications: A survey. Earth Science Informatics, 12, 143-160..
3. Burrough, P. A., McDonnell, R. A., & Lloyd, C. D. (2015). Principles of geographicalinformation systems. Oxford University Press, USA.
4. Campbell, James B., and Randolph H. Wynne. Introduction to remote sensing. Guilford press, 2011.
5. Jensen, J. R. (2009). Remote sensing of the environment: An earth resource perspective 2/e. Pearson Education India.
6. Ji, Z., Zhu, Y., Pan, Y., Zhu, X., & Zheng, X. (2022). Large-Scale Extraction and Mapping of Small Surface Water Bodies Based on Very High-Spatial-Resolution Satellite Images: A Case Study in Beijing, China. Water, 14(18), 2889.
7. Lillesand, T., Kiefer, R. W., & Chipman, J. (2015). Remote sensing and image interpretation. John Wiley & Sons.
8. Longley, P. A., Goodchild, M. F., Maguire, D. J., & Rhind, D. W. (2015). Geographicinformation science and systems. John Wiley & Sons.
9. Pettorelli, N., Laurance, W. F., O'Brien, T. G., Wegmann, M., Nagendra, H., & Turner, W. (2014). Satellite remote sensing for applied ecologists: opportunities and challenges. Journal of Applied Ecology, 51(4), 839-848.
10. Roy, B., & Bari, E. (2022). Examining the relationship between land surface temperature and landscape features using spectral indices with Google Earth Engine. Heliyon, 8(9).
11. Turner, W., Rondinini, C., Pettorelli, N., Mora, B., Leidner, A. K., Szantoi, Z., ... & Woodcock, C. (2015). Free and open-access satellite data are key to biodiversity conservation. Biological Conservation, 182, 173-176.
12. Xia, H., Zhao, J., Qin, Y., Yang, J., Cui, Y., Song, H., ... & Meng, Q. (2019). Changesin water surface area during 1989–2017 in the Huai River Basin using Landsat data and Google earth engine. Remote Sensing, 11(15), 1824.

13. Xu, H. (2006). Modification of normalised difference water index (NDWI) to enhanceopen water features in remotely sensed imagery. International journal of remote sensing, 27(14), 3025-3033.

14. Zhang, F., Tiyip, T., Johnson, V. C., Wang, J., & Nurmemet, I. (2016). Improved water extraction using Landsat TM/ETM+ images in Ebinur Lake, Xinjiang, China. Remote Sensing Applications: Society and Environment, 4, 109-118.

CHAPTER 5

MAPPING OF INLAND AND COSTAL WATER QUALITY

Vidya Shree Bharti[1] and Kantarajan G[2]

1 ICAR-Central Institute of fisheries Education, Mumbai-400061
2 ICAR-National Bureau of Fish Genetic Resources, Lucknow – 226002, Uttar Pradesh

---

Introduction

Chlorophyll a is a pigment found in all photosynthetic organisms, and is used as a measure of primary productivity in the oceans and inland waters. It can be measured indirectly through satellite remote sensing and it's calculated using the raw data coming from the satellite sensors and specific algorithms to process that data. Satellite remote sensing is a highly useful tool for monitoring chlorophyll-a concentration (Chl-a) in water bodies. Remote sensing algorithms based on near-infrared-red (NIR-red) wavelengths have demonstrated great potential for retrieving Chl-a in inland waters. With satellite remote sensing, the state of inland waters can be monitored through synoptic observations collected at frequent intervals. Generally, remote sensing of inland waters is much more difficult than that of open oceans. This is because the optical properties of inland waters are not determined only by phytoplankton (like open oceans), but are also strongly influenced by other constituents (i.e., non-algal particles [NAP] and colored dissolved organic matter [CDOM]). As a result, inland waters have more complex optical properties. Both NAP and CDOM have larger absorptions at the blue and green spectral regions, and thus make these bands inappropriate for retrieving chlorophyll-a concentration (Chl-a) in many inland waters.

Optical properties of water

Light entering inside a water body is either absorbed or scattered by the sea water constituents. The amount of light that is absorbed gives the absorption coefficient and the amount of light that is scattered gives the scattering coefficient. Since these properties solely depend on the type and the concentration of sea water constituents, they are termed as Inherent Optical Properties (IOPs). The optical properties that depend on the IOPs and the ambient light field are called Apparent Optical Properties (AOPs). They include both scalar and vector irradiance, radiance and other optical properties that can be derived from irradiance and radiance such as remote sensing reflectance (Rrs) and diffuse attenuation coefficient (K). These AOPs and IOPs together describe the optical behavior of the water body. They are inter-related and vary with change in concentration of in-water constituents.

Optical classification of oceanic and Inland waters

Morel and Prieur (1977), Gordon and Morel (1983) classified oceanic waters as Case 1 and Case 2 waters. Case 1 waters such as open ocean waters where the optical properties are primarily controlled by the phytoplankton are much easier to understand than the optically complex case 2 waters where the optical properties are controlled not just by the phytoplankton, but also by other organic and inorganic substances that vary independently ofphytoplankton. Thus, lake, estuaries and coastal waters are often referred as case 2 waters. Due to the complex interaction between the physical, chemical and biological processes and the lack of knowledge on optical properties it is more difficult to interpret the signals received from the case 2 waters than from case 1 waters. Most of the standard ocean color algorithms used for estimating the concentration of sea water constituents provides poor results when applied to case 2 waters than case 1 waters. In order to improve the accuracy of estimating sea water constituents of coastal and inland case 2 waters, it is required to better understand the bio-optical properties of coastal waters, their interrelationships and their relationship with sea water constituents, their variations in time and space etc.

Ocean color remote sensing

Understanding of the inter relationships between the various optical properties and the in- water constituents helps in developing effective bio-optical algorithms that will extract the information on concentration and the type of water constituents. The various optical properties can now be measured using the state-of-the art optical instruments. However, under certain circumstances such as lack of manpower, sea surface conditions, financial limitations in procuring instruments, non-existence of instruments it is always not possible to have a complete set of optical properties. These missing optical properties are often derived from the available ones using either radiative transfer models or empirical algorithms. Since the data generated by radiative transfer models does not depend on any empirical relationships and is expected to be free from measurement errors (Mobley et al. 1993), it can be efficiently used for developing bio-optical empirical algorithms for extracting the information from remotely sensed images. However, it is always advisable to carry out the process of optical closure where the modeled and the measured optical properties are compared against each other to ensure the quality of the data generated using radiative transfer simulations.

For case 1 waters, since the optical properties are only controlled by phytoplankton and its derivatives, the in-water constituent's concentration can be directly related to the remotely sensed signals. Whereas, for case 2 waters, due to their complexity, it is sometimes not possible to examine which water constituent is actually responsible for modifying the spectral shape of the remotely sensed signals, thus increasing the difficulty level of developing the bio-optical algorithms. Also, since the coastal waters or case 2 waters are influenced by the land activities, they have variable physical, chemical, biological, geological and thus optical properties. Therefore, extracting information from remote sensing signals requires season and region-specific algorithms.

Monitoring open, coastal oceans and Inland water

The traditional methods of monitoring open and coastal ocean using ship cruises and buoys have limited spatial and temporal extent. With the advent of ocean color remote sensing satellites along with the online availability of satellite data, monitoring open and coastal ocean over a wide synoptic range has become an easy task. The fact that the amount and typeof constituents of water changes the optical behavior of the water body, thus giving different color to the water body is used as a physical basis for the development of ocean color satellite sensors (Alberotanza 1989). The remote sensor observes the radiance emerging from the water body, thus providing primary information on ocean color. This data remains totally useless to the public unless it is related to some phenomena occurring in the ocean. The ocean color satellite data can be made useful to the public by developing appropriate empirical algorithms to retrieve the products of public interest. Ocean color remote sensor observes the emerging radiance from water that carries primary information on the ocean color. As the light in ocean is primarily controlled by the amount and the type of biological substances present in it, these properties are often termed as bio-optical properties and the algorithms developed for interpreting the ocean color satellite data are called as bio-optical algorithms. The two types of algorithms are semi analytical and empirical bio optical algorithms. Semi analytical models are those that depend on the actual relationship between two optical parameters derived using radiative transfer models. These models are based on various assumptions such as sea state condition, surface illumination, spectral shapes of optical parameters etc.

Empirical algorithms provide direct relationship between two optical parameters. For example, either ratios or the absolute values of remote sensing reflectanceare related to the other optical property or directly to sea water constituents such as chlorophyll, sediments etc.

Chlorophyll estimation in coastal water

The surface Chl concentrations can be estimated by using NASA Ocean Chlorophyll 4 (OC4) bio-optical algorithm within SeaDAS software. The SeaWiFS OC4 algorithm is a maximum band ratio algorithm employing 4 spectral bands: band 2, 3, 4 and 5. Band 5 (555 nm) is taken as the reference wavelength band and three band ratios are computed for each measurement of the radiance or reflectance spectrum as given in O'Reilly et al. (1998):

$$\text{Chl-a} = 10^{(0:366\ 3:067R\ +1:930R2\ +\ 0:649R3\ 1:532R4\ )}$$

where $R = \log_{10\ [\{}$ Rrs(443) > Rrs (490) > Rrs (510) }/ Rrs(555)]

Rrs is the remote sensing reflectance. The OC4 algorithm is purely empirical in that it applies simple regression between the field determinations of pigment concentration and spectral ratios of ocean reflectance or normalized water-leaving radiance. The OC4 algorithm has been considered as standard method for satellite detection of HABs over the global waters.

Chlorophyll estimation in inland water

Regression model are developed with remote sensing reflectance and in-situ chlorophyllconcentration and after validation of the model they are applied for chlorophyll estimation.

Model for TSI-chl-a during post monsoon (October to January) Chl-a = -11.833 + 992.816 Rrs_{B2}

Model for Chl-a during pre - monsoon (February to May) Chl-a = -100.660 + 107.171 $Rrs_{B3/B1}$

Chl-a = -70.942 + 100.101 $Rrs_{B3/B1}$ – 370.877 Rrs_{B2}

Where:

$Rrs_{B3/B1\ -}$ Reflectance ratio of NIR and green band for the station concerned

$Rrs_{B2\ -}$ Reflectance in the red band for the station concerned

Rrs_{B1} - Reflectance in the green band for the station concerned

Rrs_{B3} - Reflectance in the NIR band for the station concerned

Suggested Readings

1. Carder, K. L., R. G. Steward G. R. Harvey and P. B. Ortner 1989. Marine humic and fulvic acids: their effects on remote sensing of ocean chlorophyll. Limnology and Oceanography. 34:68-81.
2. European Environment Agency (1998) European Environment Agency (1998) Report.
3. Jerlov, N.G., 1976. Marine Optics. Elsevier, New York (231 pp.).
4. Kirk, J. T. O. 1994 Light and Photosynthesis in Aquatic Ecosystem. Cambridge University Press, Bridge.
5. Longley, P. A., Goodchild, M. F., Maguire, D. J., & Rhind, D. W. (2015). Geographic information science and systems. John Wiley & Sons.
6. Pettorelli, N., Laurance, W. F., O'Brien, T. G., Wegmann, M., Nagendra, H., & Turner, W. (2014). Satellite remote sensing for applied ecologists: opportunities and challenges. Journal of Applied Ecology, 51(4), 839-848.

CHAPTER 6

REGISTRATION AND ACQUISITION OF FREE SATELLITE DATA FROM BHUVAN, USGS, ESA, ACQUIRING TOPOSHEETS FROM SOI, AND ORDERING OF IRS DATA

Vinod Kumar Yadav, Mahesh Sharma, Liton Paul, and Akhil Sambhe

ICAR-Central Institute of fisheries Education, Mumbai-400061

What you should be able to do after this practical?

1. You will be able to know about Bhuvan, USGS, ESA, and IRS satellite data sources andSOI toposheets.
2. You will be able to make personal account and search for the dataset.
3. You will be able to save results/download/order/make request/ for the dataset.

What is the Bhuvan Indian geo-platform of ISRO? What kind of information canyou get from there?

Bhuvan, is the national Geo-portal developed and hosted by ISRO comprising of Geo SpatialData, Services and Tools for Analysis. It has many versatile features such as:

1. Visualization of Satellite Imagery and Maps
2. Analysis
3. Free Data Download and
4. Download Reports to name a few.

The Satellite Imageries are of multi-sensor, multi-platform and multi-temporal in nature canbe visualized in 2D and 3D. Varieties of thematic maps are also hosted for analysis of variousnatural resource themes. Bhuvan platform supports many applications that address Governance andother Geo Spatial applications that are being used by the Central and State GovernmentsDepartments, Academia and Industry.

What is USGS Earth Explorer? What kind of information can you get from it?

USGS Earth explorer is an online tool that serves as an interface to access earth science data from archives of USGS Earth Explorer provides online search, browse, display, metadataexport and data download from the archives of the USGS (United State Geological Survey).The Archive spans from 1937 aerial photographs to millions of satellite images of theEarth's surface, starting with the original Earth orbits in the 1960's and first Landsat satellite in 1972, to their current hourly additions of satellite images. The USGS Earth Explorerproducts include the following:

- Aerial Photography
- Advanced Very High-Resolution Radiometer (AVHRR)
- CEOS Legacy
- Commercial Satellites
- Declassified Data
- Digital Elevation
- Digital Line Graphs (DLG)
- Digital Maps
- Earth Observing One (EO-1)
- Global Fiducials
- Global Land Survey (GLS)
- Heat Capacity Mapping Mission (HCMM)

- ISERV
- ISRO Resourcesat
- Land Cover Products
- Landsat Satellite Archives
- Landsat Satellite Legacy
- NASA LPDAAC Collections
- Radar
- Sentinel-2
- Unmanned Aircraft Systems (UAS)
- Vegetation Monitoring

What is ESA open access Copernicus hub and what information it gives?

The Copernicus Open Access Hub (previously known as Sentinels Scientific Data Hub) provides complete, free and open access to Sentinel-1, Sentinel-2, Sentinel-3 and Sentinel-5Puser products. The open hub tool of its website provides sentinel 1, 2 and 3 data. The European Space Agency is an intergovernmental organization of 22 member states dedicatedto the exploration of space. Established in 1975 and headquartered in Paris.

What is a Toposheet? Where can you download one?

In modern mapping, a topographic map or topographic sheet is a type of map characterized by large-scale detail and quantitative representation of relief features. A topographic survey is typically based upon systematic observation and published as a map series, made up of two or more map sheets that combine to form the whole map. A topographic map series uses a common specification that includes the range of cartographic symbols employed, as well asa standard geodetic framework that defines the map projection, coordinate system, ellipsoid and geodetic datum. We can download an Indian region toposheet or a Mapsheet from Survey Of India –Online maps Portal|| for free.

What do you know about IRS?

Indian Remote Sensing Missions (IRS) observes Earth with Optical, microwave and hyperspectral instruments flown on-board to provide necessary data. NRSC (National Remote Sensing Centre) is the nodal centre for hosting Satellite Data Products from more than 13 IRS satellites right from the first IRS optical mission namely IRS-1A and SAR imaging missions. NRSC Data archive is extensively utilized for Land use land cover monitoring, Ocean studies, weather applications and Scientific research. The Government agencies, industries and academia is highly benefiting by the valuable huge data products archive and technology to meet their respective end objectives.

Procedure

Bhuvan data access

Step 1: Registration and sign in:

- Go to Open Data Archives on the home page of Bhuvan website. On the top rightcorner, click on the -login|| option.
- The Bhuvan-Single Sign On box will appear.
- Click on -New User?|| option.
- Fill the required mandatory details to create your account in Account and profileinformation form.
- Click -Submit||.
- You will soon receive an activation link in your registered email shortly aftersubmitting.
- Check your mailbox of registered mail id and click on the link to completeregistration process.
- An Activation Page appears. Fill to create and confirm a password for your account.
- The Activation Page will confirm your registration with informing you about the logincredentials.
- Visit the Bhuvan open data archives to -login|| again with these credentials.

Step 2: Applying Search:

- There are 3 selection categories
  1. Satellite/sensor
  2. Theme/Products
  3. Program/Projects
- Select satellite/sensor and select any subcategory for selecting specific sensor in thelist.
- For this example, we will search a -Resourcesat-1/Resourcesat-2 LISS III dataset.

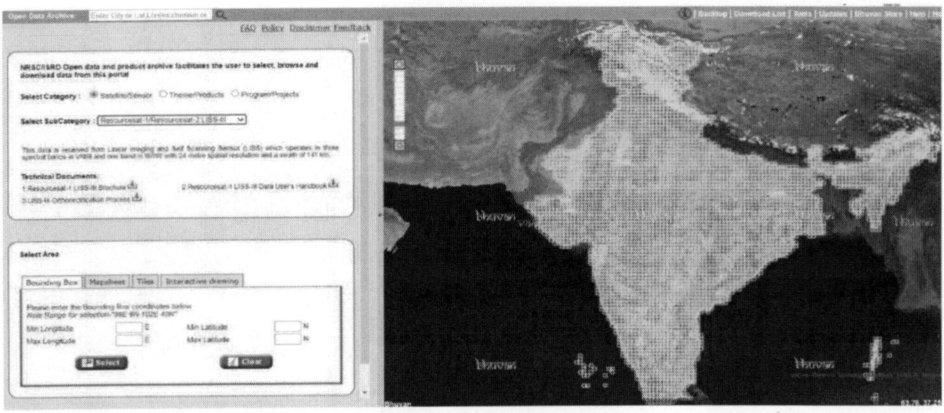

Figure 6.1: ISRO- Bhuvan

- In this lab, first we will search data with the -TILE method.
- For example, zoom in to the Andaman region. Here you can see tiles shown
- Now in the select area box, click -Tiles and click -start|| to select a tile.
- Press Shift + Click to select multiple tiles. Click stop to stop selection and then click NEXT.
- You will get the dataset list for the selected tiles as shown below:

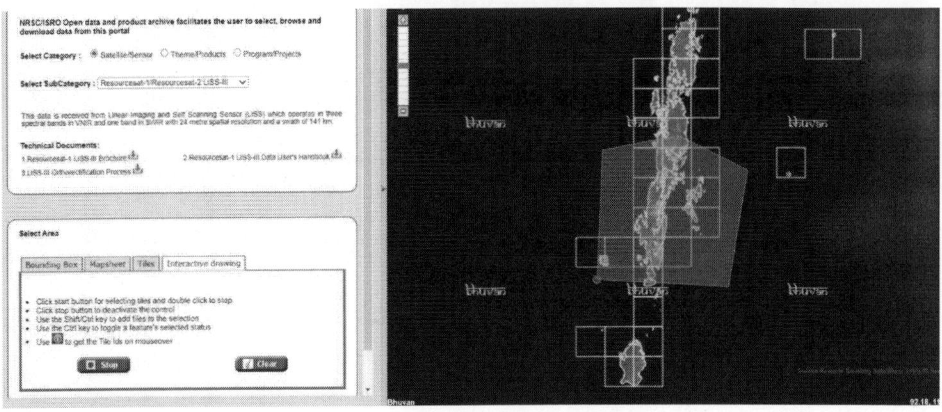

Figure 6.2: Search data by tiles

- In the method of interactive drawing, Click start and click on to make a polygon toselect the tiles present in that polygon.
- Double click to complete the polygon.
- Click Stop to stop the selection and click Next.
- Again, a list of datasets will appear.

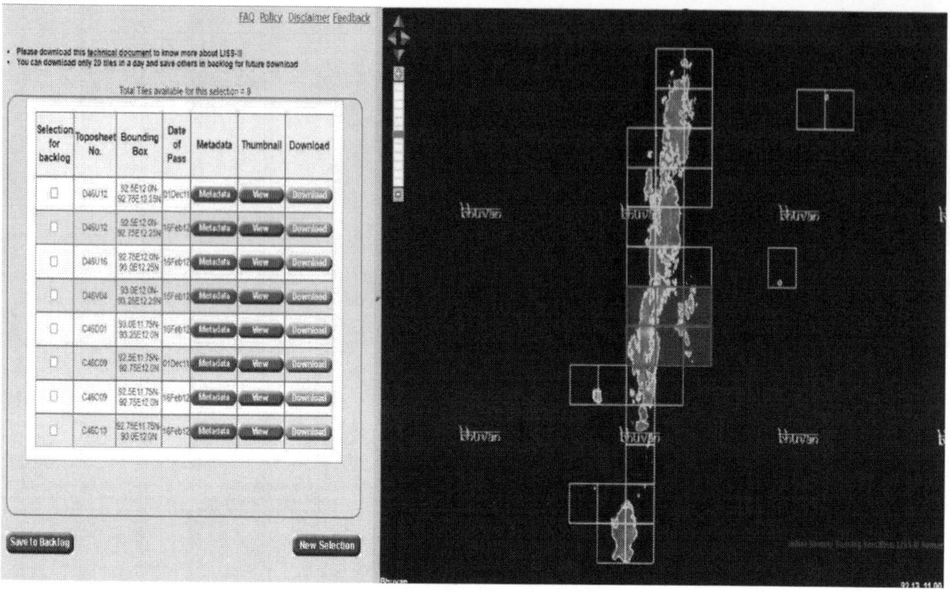

Figure 6.3: Download the data

Step 3: Saving Results

- You can preview the thumbnail view by clicking -View||.
- Finally click -Download|| in the Total tiles available list and your download will start.
- You will receive a zip file.

USGS Earth Explorer data access

Step 1: Registering and sign in

- There is Login, Help and Feedback option on the Earth Explorer user interface menubar.
- Click to Login option to begin
- Click create new account by entering Username and confirm new password tocontinue.
- Next enter the Contact demographic information. After submitting the details, a review of your details will be displayed. Click Submit Registration for finally registering to the online tool.
- An account confirmation link will be sent to your registered email address. Click that link and confirm your username sent in mail. Return to login page and sign in with theprevious credentials. Your profile home page will appear.

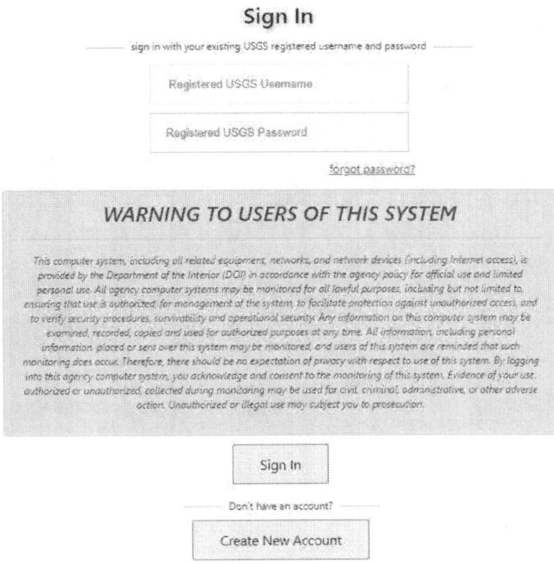

Figure 6.4: USGS registration

Step 2: Search Methods

- Go to https://earthexplorer.usgs.gov/
- In the search criteria, type in an address or place name, enter coordinates or click the map to define your search area.
- For example, in the Geocoder we select Address/Place and type New Delhi and click show. Click on New Delhi in Address/Place to display the location on the map.

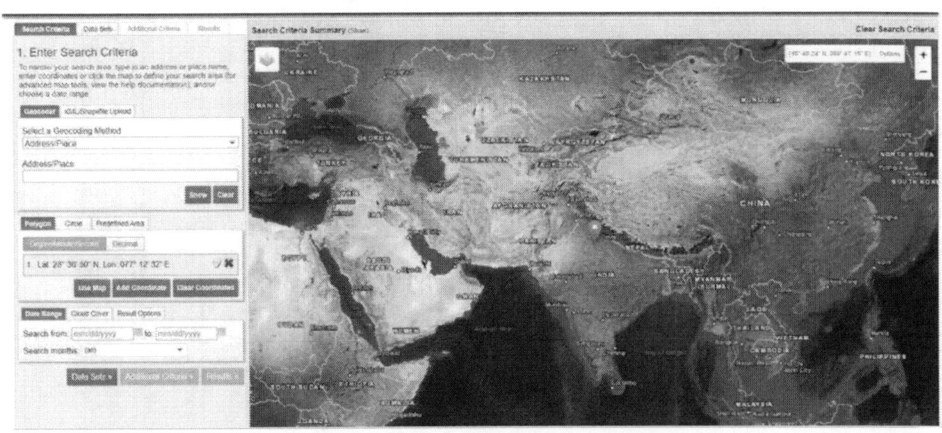

Figure 6.5: Search the location

- The Worldwide Reference System (WRS) is used to identify the path and row of each Landsat image. The path is the descending orbit of the satellite. Each path is segmented into 119 rows, from north to south.

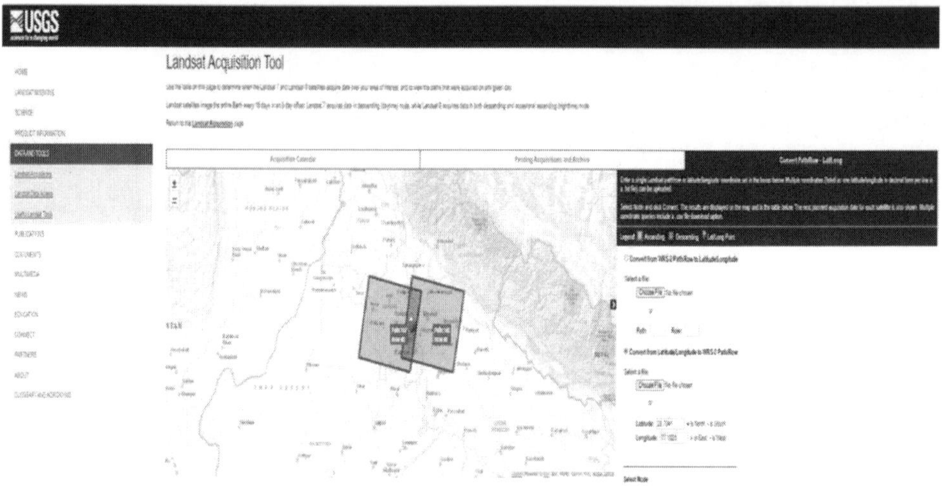

Figure 6.6: Search the location in USGS using path and raw

- WRS is a global notation used in cataloging Landsat data. Landsat satellites 1, 2 and 3 followed WRS-1, and Landsat satellites 4,5,7, and 8 follow WRS-2.
- Search google for the LANDSAT world reference system path row.
- Navigate to https://landsat.usgs.gov/landsat_acq#convertPathRo LANDSATAcquisition Tool.
- For example, enter the Latitude and longitude of New Delhi and select convert to Pathand Row. Yow will get the desired path as 146 and row as 40.
- Now enter these details in USGS for the path/row search criteria.
- For feature selection, enter the geographic feature name as New Delhi, select country,select a feature class from list and a feature type.
- Click on the Place name. It will be displayed on the map interphase
- After that select Polygon and then select add coordinates in the Degree/ minute/secondoption. Enter the lat and long details.
- Select use map option to select points to enclose a polygon

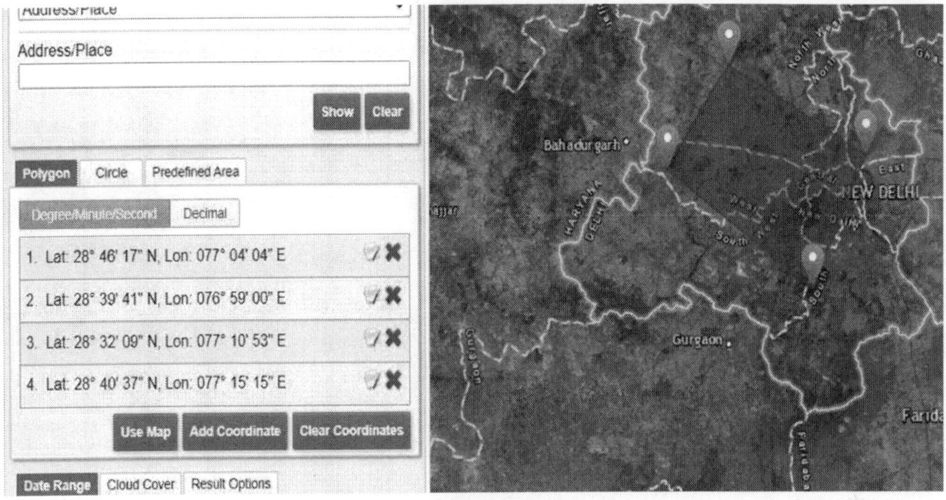

Figure 6.7: Select the polygone using points

- Enter the date range and click Data sets.
- In the Datasets, select the data source you want to have for example NAIP aerialimagery or Landsat 8 OLI/TIRS L2 data
- Click additional criteria to enter additional details like row path, cloud cover, daytime or night time imagery etc.
- Finally click results.

B. In the results pane in the left side of the page you can see footprint, download, quick-view and add to basket options for each image result.

- For example, a quick-view preview is overlayed is displayed on area.

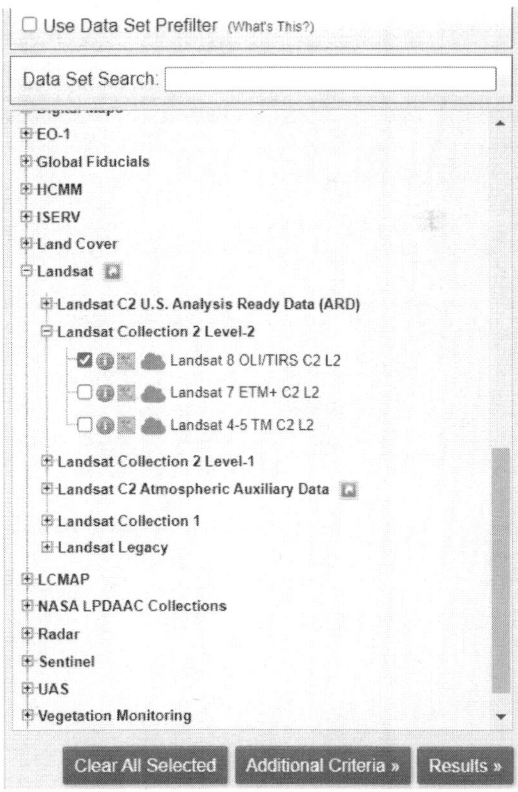

Figure 6.8: Select the dataset

Figure 6.8: Select the dataset

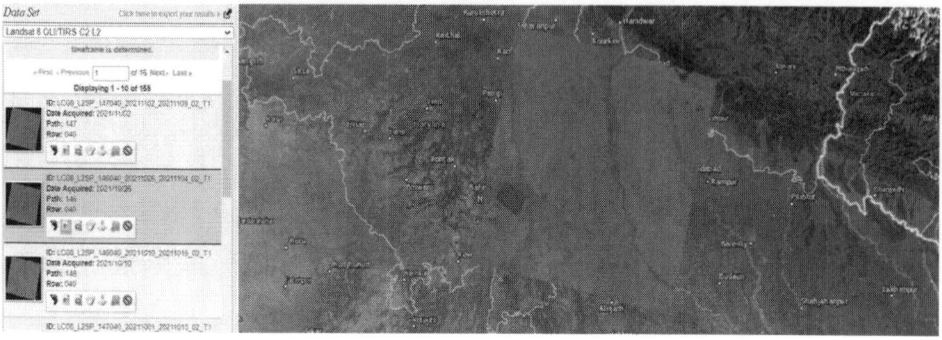

Figure 6.9: Available dataset

Step 3: Saving Results

- Click the download icon and select the product required.
- You can download the individual related file or can download the whole dataset. Azip file will be downloaded for full dataset.

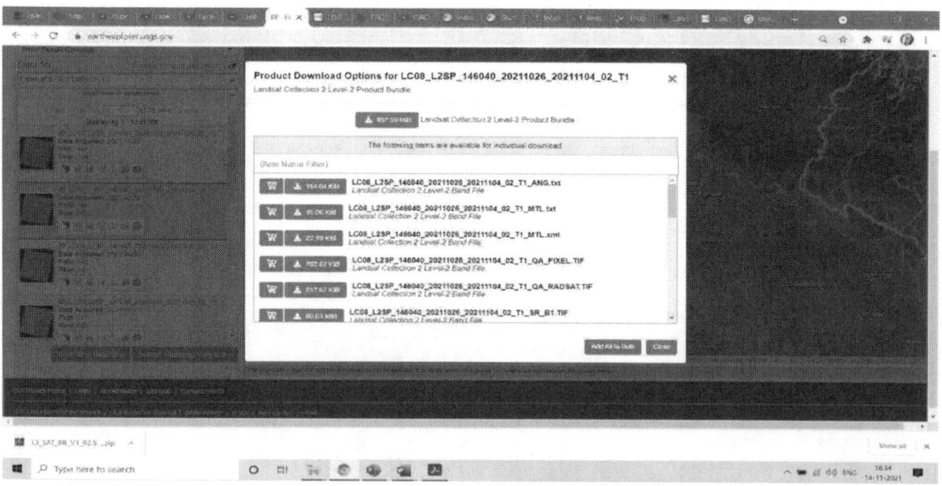

Figure 6.10: Download the data

ESA data access

Step 1: Registration and sign in

- Go to the https://scihub.copernicus.eu/ Copernicus open access hub. And click to theOpen Hub component on the page. It will direct yo to the catalogue.
- The open hub provides sentinel 1, 2 and 3 data. In order to download the

data, youneed to sign in or sign up.

- Click sign up and register for an account by filling the details required. Click register.
- After login you can perform search as well as download dataset.

Figure 6.11: ESA registration

Step 2: Applying Search:

- Navigate to Mumbai Area. Switch to the navigation mode in the right side to select a polygon of desired area.
- Let's search for a Sentinel 2A data of the Mumbai region.
- Enter the sensing period range of interest in the Advanced search option.
- You can define the product type and give a cloud cover percentage.

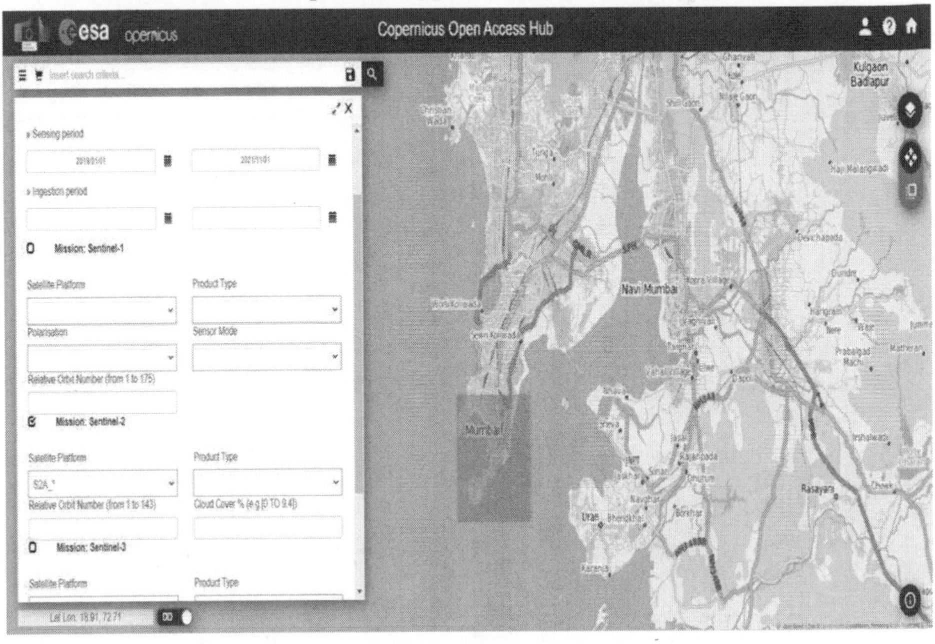

Figure 6.12: Selection of the area using polygone

- Use a polygon you have drawn in earlier step.
- And click the lens (search) icon to get desired list of datasets available.

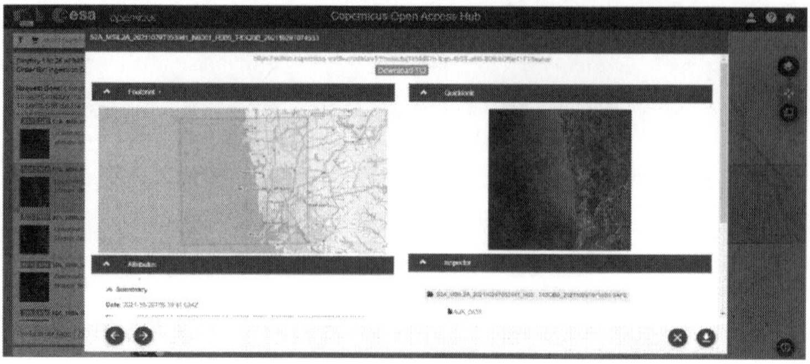

Figure 6.13: Quick look of the dataset before downloading

Step 3: Saving Results:

- A quick look can be generated using the eye icon in a dataset from the list. From thereyou can download the product and also can summarize the product. You can get file list of the dataset.
- A zip file download gets started having all the files relevant to your search. Alternatively, you can download from the Download product icon from the main lists of products

SOI toposheet access

Step 1: Registration and sign in

- Go to Survey of India website https://www.surveyofindia.gov.in/ and click on the map services. And select Online maps portal (Free/Purchase).
- Click sign in at the top right corner and follow the steps after clicking Register Here option.

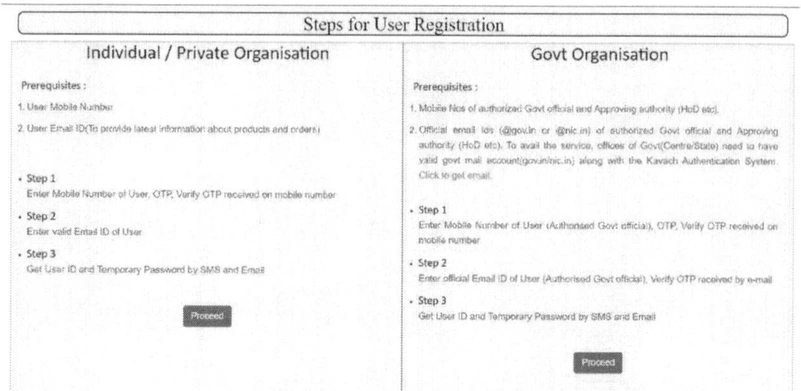

Figure 6.14: Survey of India registration

- You will be required to give and verify your mobile number and email id details.
- Your login credentials User ID and Temporary Password have sent to your MobileNumber and Email ID.
- You have to change your password when login first time.

Step 2: Applying Search

- Go to the Products and select Open series map (FREE PDF).

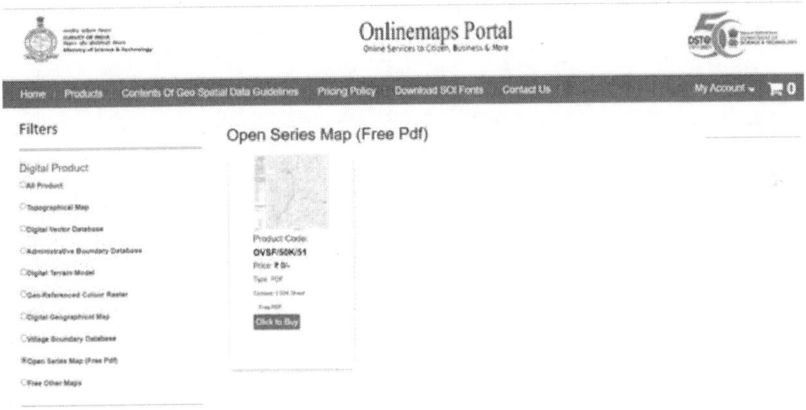

Figure 6.15: Survey of India online maps portal

- Now you will be required to enter the SHEET NUMBER.
- For this, click SEARCH SHEET NUMBER.Select state as Delhi and district as New Delhi for example. Click Select to select thesheet number.

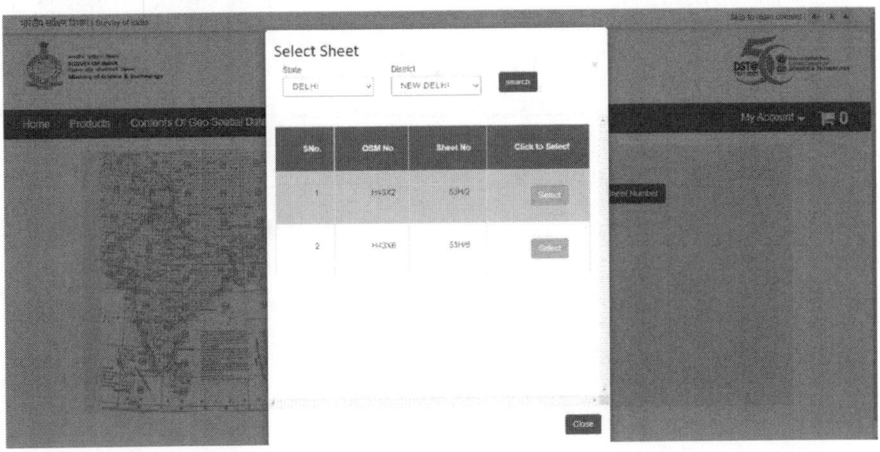

Figure 6.16: Enter the state and district names to get the sheet number

Step 3: Saving Results

- Click Download. A toposheet with a scale of 1:50000 will get downloaded in a pdf file format.

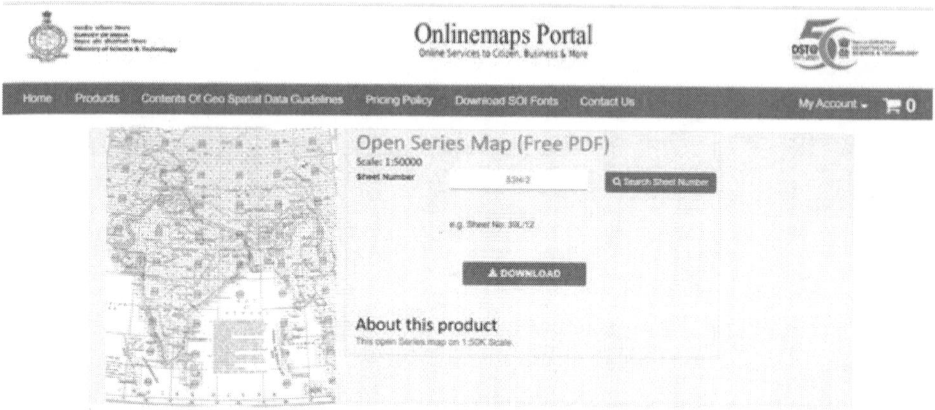

Figure 6.17: Saving toposheet

E. IRS dataset request

Step 1: Registration

- Go to https://www.nrsc.gov.in/EOP_irsdata_Links
- Click on the Data Dissemination Portals for the data product of the Indian Missions.
- Click User order portal

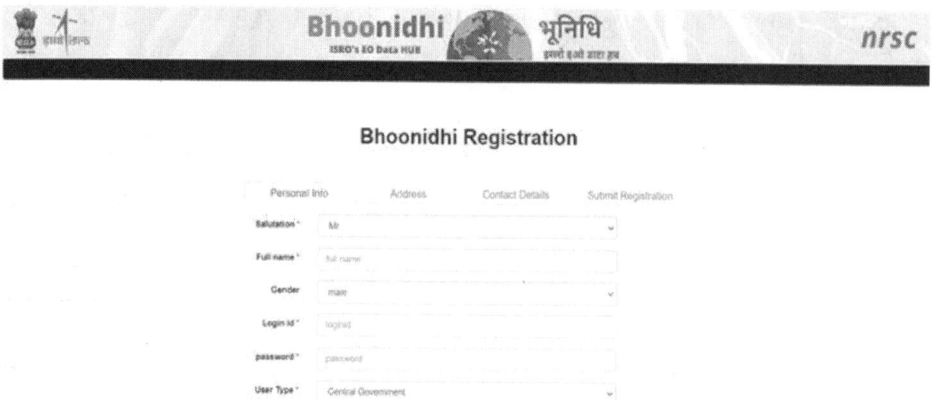

Figure 6.18: Bhoonidhi registration

- Go to New Bhoonidhi Version: https://bhoonidhi.nrsc.gov.in
- Apply a search for example Mumbai area and select a Priced data.
- There you can see ResourceSat AWIFS mission data archives.

- You will need to register by filling details in the Bhoonidhi Registration form. And verify through the link you receive in your email id after submitting.

Step 2: Searching Data and ordering

- Apply search by name or polygon drawing. Select a Priced sensor ResourceSat- 2 and2A for example.
- A list of products available will appear. Get the Metadata view from the side icons inthe list.

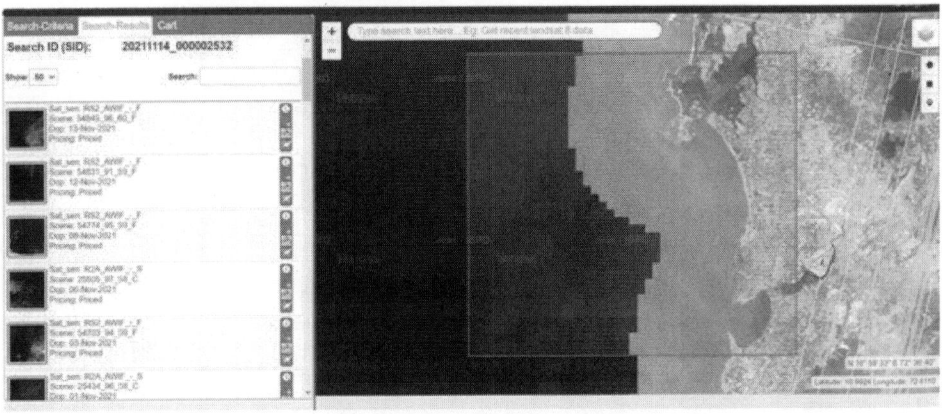

Figure 6.19: Search of resourcesat 2 and 2A data

- Click on to the add to cart iconNext click Select Product specification.
- Cost and product information is specified in there. After adding product specifications, you can fill details to purchase the data

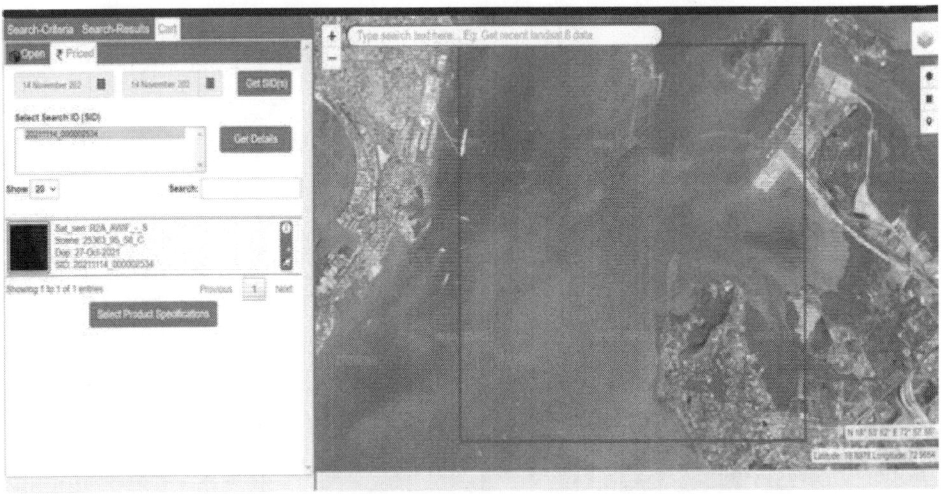

Figure 6.20: Select product specification before ordering

CHAPTER 7

SPATIAL QUERIES, GEO-REFERENCING AND MAP PREPARATION

Vinod K kumar Yadav, Palsam Kartik Kumar Goud and Liton Paul

ICAR-Central Institute of Fisheries Education, Mumbai-400061

Projection and Reprojecti on Using Quantum GIS

What you should be able to do after this exercise?

We will get familiar with the projecting and reprojecting the vector and raster data sets byusing Quantum GIS.

Principle

As we know the earth is spherical in shape but its surface is not regular at all places. It contains mountains, valleys and plains. Due to this, it is not suitable to perform direct mathematical calculations on it. Therefore, a mathematical surface of oblate spheroid best fitting the irregular surface of the earth is considered as a frame of reference for measuring locations on the surface of earth and is called as Datum. If the spheroid is best fitting a particular area, then it is called as Local Datum and if it is best fitting the entire globe then it is called as Global Datum. For example, Everest ellipsoid is the best fit to India and its adjacent countries but not to the entire world, similarly WGS 84 ellipsoid is the best fit to theentire world but not exactly to India. Generally Geographic Coordinate Systems (GCS) are associated with this datum, the measuring units are angular in nature. It is very hard to measure distance, area and other parameters by using spherical coordinates and also i t is very difficult to represent this three-dimensional curved surface on a piece of paper. To overcome this difficulty, we use a method called as map projection. It aims to convert the earth's 3D curved surface to a map's two-dimensional flat surface i.e., on a piece of paper. For this purpose, we use projection surfaces like Plain, Cone and Cylinder (see the following figure). Since, it is impossible to convert a curved surface to a flat surface accurately, we have to compromise on some properties to preserve some other properties. For example, to preserve the Shape of a feature, we have to compromise on its area and distance. Therefore, a projection is never accurate. Choosing the right projection is very important for successful GIS projects.

Procedure

Step 1: Open the QGIS desktop via, Start All Programs QGIS Dufour QGIS Desktop

Step 2: Add the vector layer Pune_District_UTM43N.shp' via., Add Vector layer button' Click on Browse' in the popup window browse to the GIS_002_Data' folder. Make sure the file type should be ESRI Shapefiles' select the Pune_District_UTM43N.shp' files Clickon Open' and again on Open'.

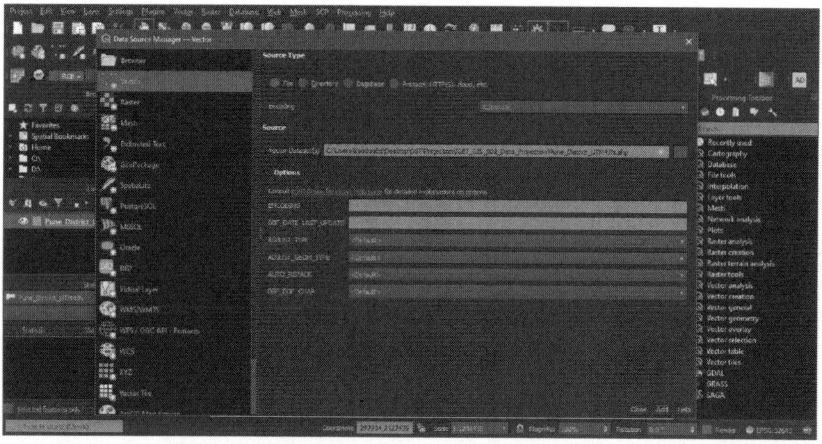

Figure 7.1: Add vector layer

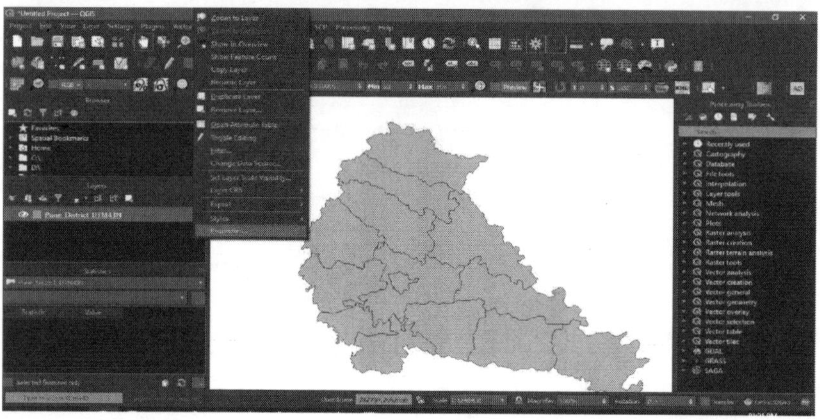

Figure 7.2: Direction to open the properties of the data

Step 3: Now Right-Click on the Pune_Disctrict_UTM43N.shp' layer under Layer' section → Select Properties.

Step 4: In the Layer Properties' popup window Click on Metadata' tab Click on Extent Tab And select the Set CRS from Layer option which will show the Coordinate/Spatial Reference System (CRS/ SRS) of the layer. The layer we selected is in UTM projection's 43rd Zone based on WGS 84 datum in northern hemisphere. Click on _OK'.

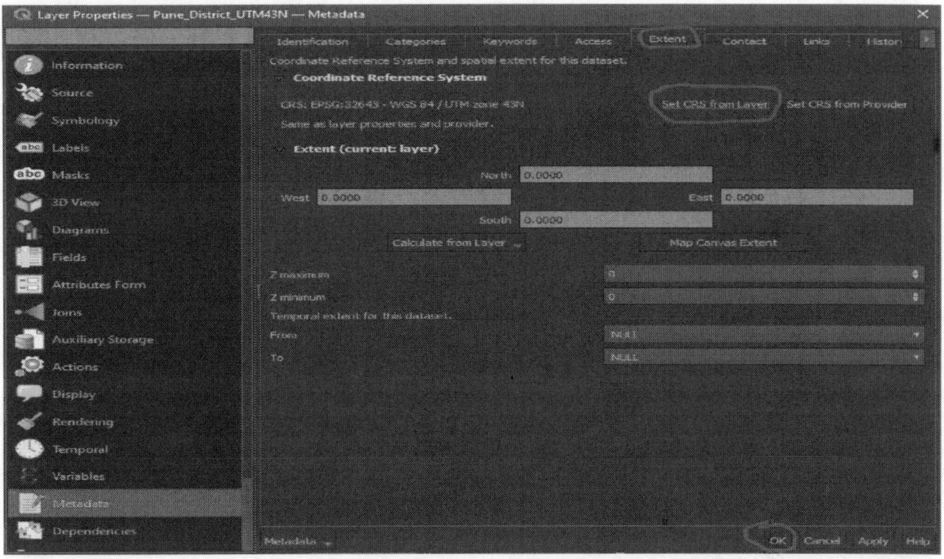

Figure 7.3: Change the projection of the data

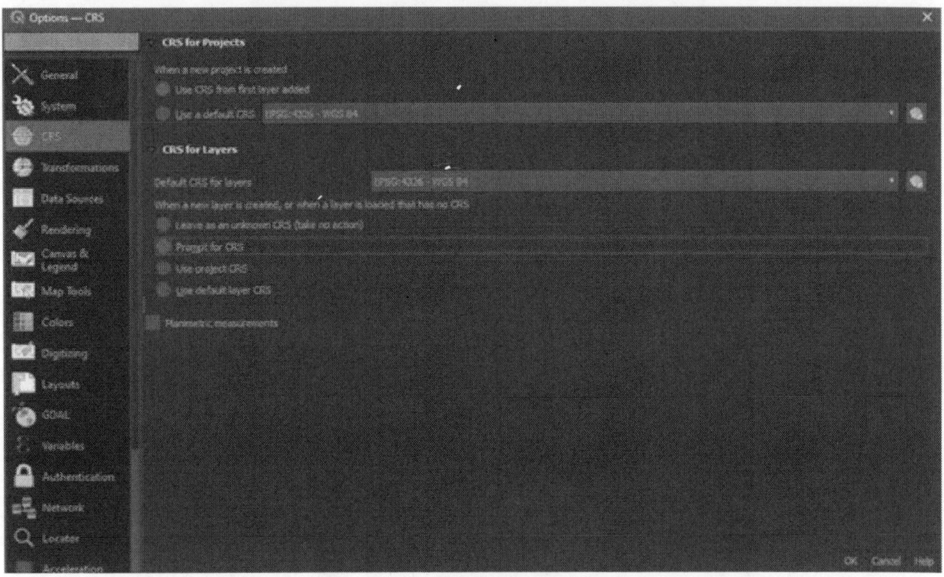

Figure 7.4: Add the coordinate reference to the data

Step 5: If the map layer does not contain any projection (CRS) information, you will be prompted to select the CRS information via, _ Coordinate Reference System Selector‘ window. Now add the vector layer _Haveli_Pune.shp‘ to the QGIS Map Canvas (refer Step2) → Now you will be prompted to select the layer CRS → select WGS84/ UTM zone 43N from_ coordinate reference systems of the world‘ section or you can use the Filter Options tofind the projection easily → Click on _OK‘. Now you can see Haveli‘ taluka added to the right place on Map.

Note: This Prompt will work only if the radio button set to it under CRS tab. This can be accessed via, Main menu bar → Settings → Options → CRS tab (see below figure). It is advised to set the radio button to Prompt for CRS‘ by default to avoid difficulties.

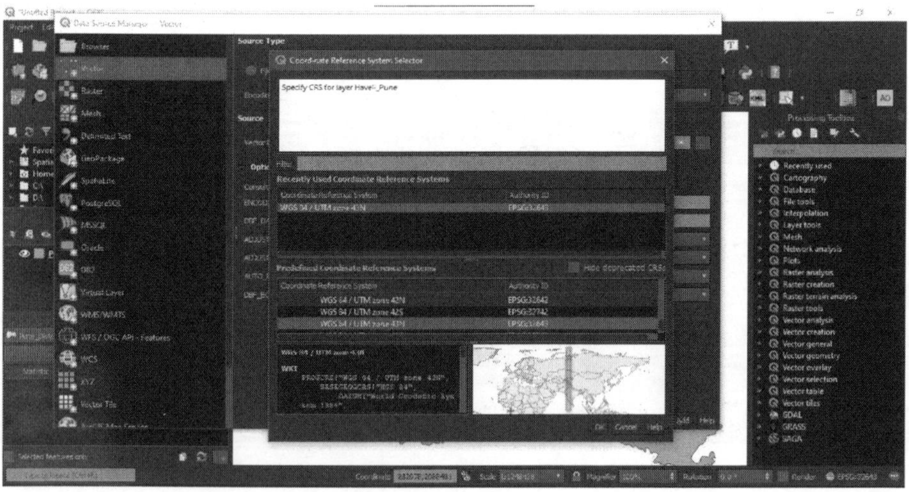

Figure 7.5: Coordinate reference system selection.

Reprojecting a Vector layer

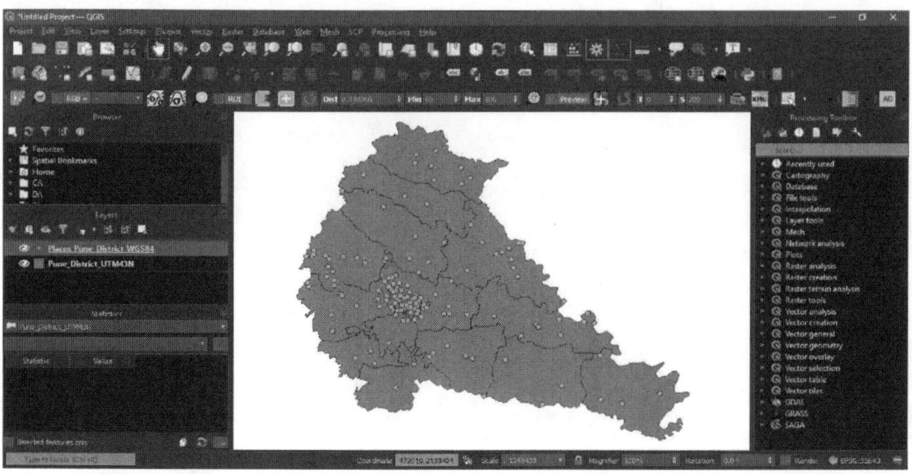

Figure 7.6: Step 6, Add another vector layer – "Places_Pune_District"

Reprojection of vector layers from Geographic to the projected system is often required for various geoprocessing analyses. On the fly helps to visualize the layer on right location but doesn't alter the projection information of the layers. In this section, we will learn how to reproject a shapefile in WGS84 (i.e., Geographic Coordinate System) containing place names of Pune district to the UTM projection.

Step 6: Add Places_Pune_District_WGS84.shp' vector layer from data using (refer Step 2).

Step 7: Check the projection information refer Step 4, you can see the layer is in GCS: WGS84.

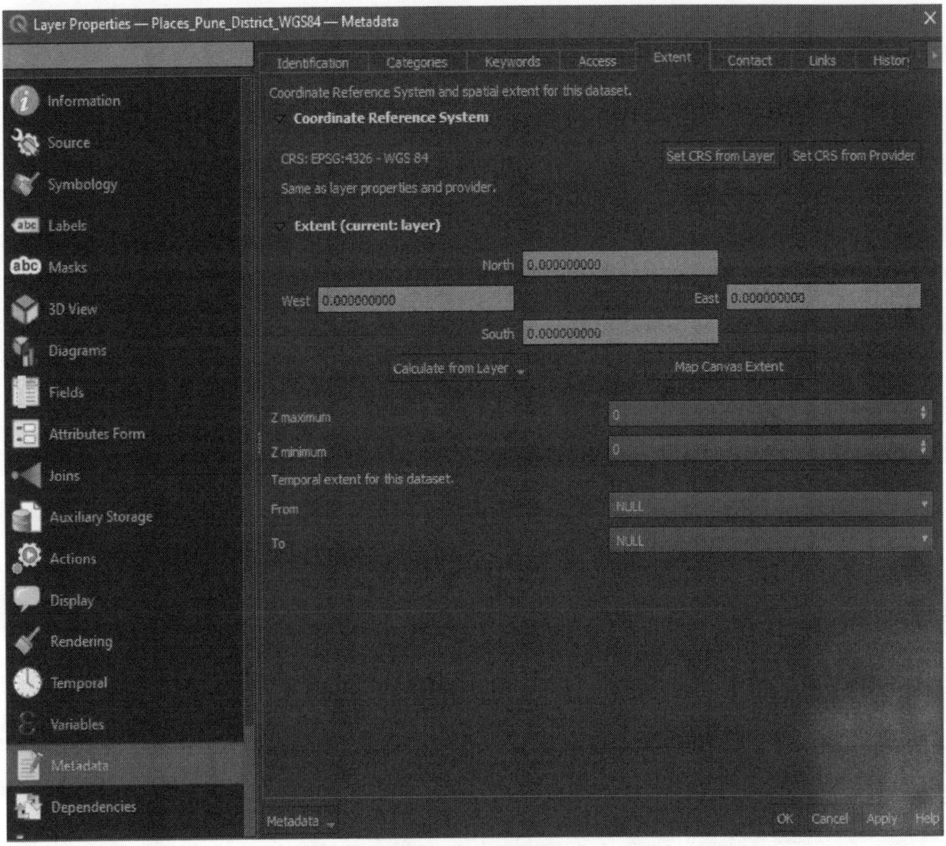

Figure 7.7: Step 7, Check the projection information

Step 8: Now Right-Click the Places_Pune_District_WGS84.shp' under Layers' selection.Click on Export' then select Save Feature As'.

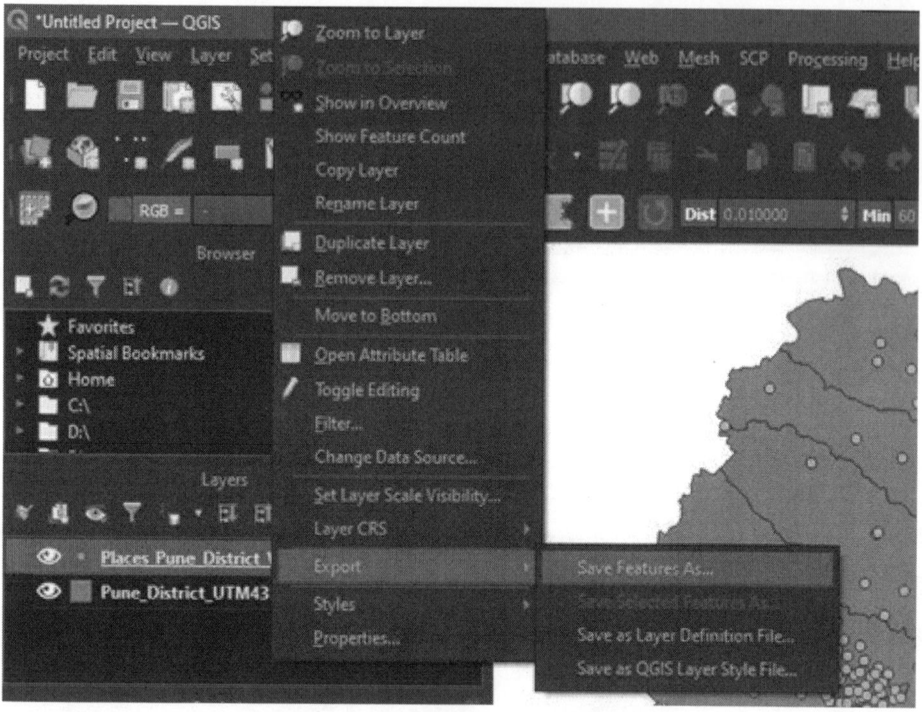

Figure 7.8: Step 8, Export features

Step 9: In _Save vector layer as‘ window select desired format you want, in this case _ ESRI Shapefile‘ → click on _Browse‘ right side to _Save As‘ → Navigate to the desired path that you want to save file. Name it as _Places_Pune_District_UTM43N.shp‘ → Click on _Save‘ → Under _CRS‘ section select, _Selected

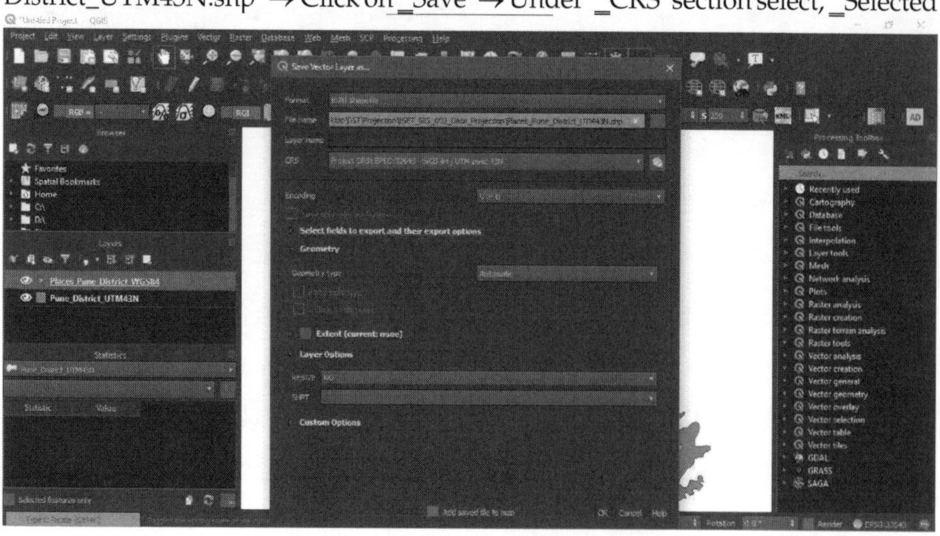

Figure 7.9: Step 9, Save vector layer

CRS' from the down drop list → Now you can see CRS selection box will be enabled, click on → _Browse' Use filter to select the WGS84/ UTM zone 43N in the popup window (refer Step 5) → Click _OK' → _Check' the check box of Add saved file to map' → Click on _OK'.

Step 10: You can notice the newly added layer under Map Layers section.

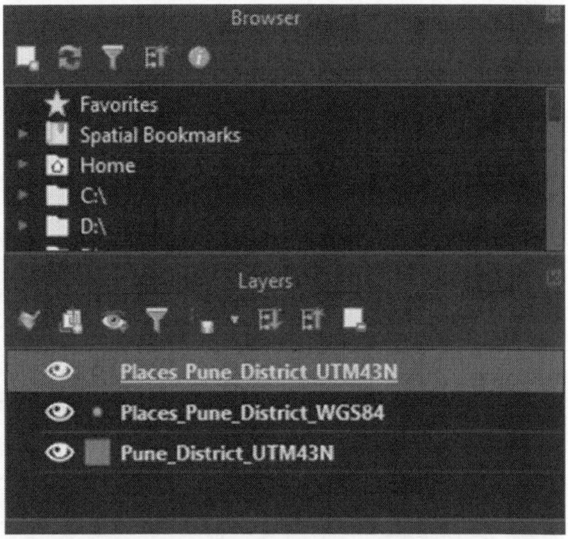

Figure 7.10: Step 10, Newly projected layer added on the map layer section

Step 11: Now check the _Metadata' tab for information about CRS of _Places_Pune_ District_UTM43N.shp', it should look like below

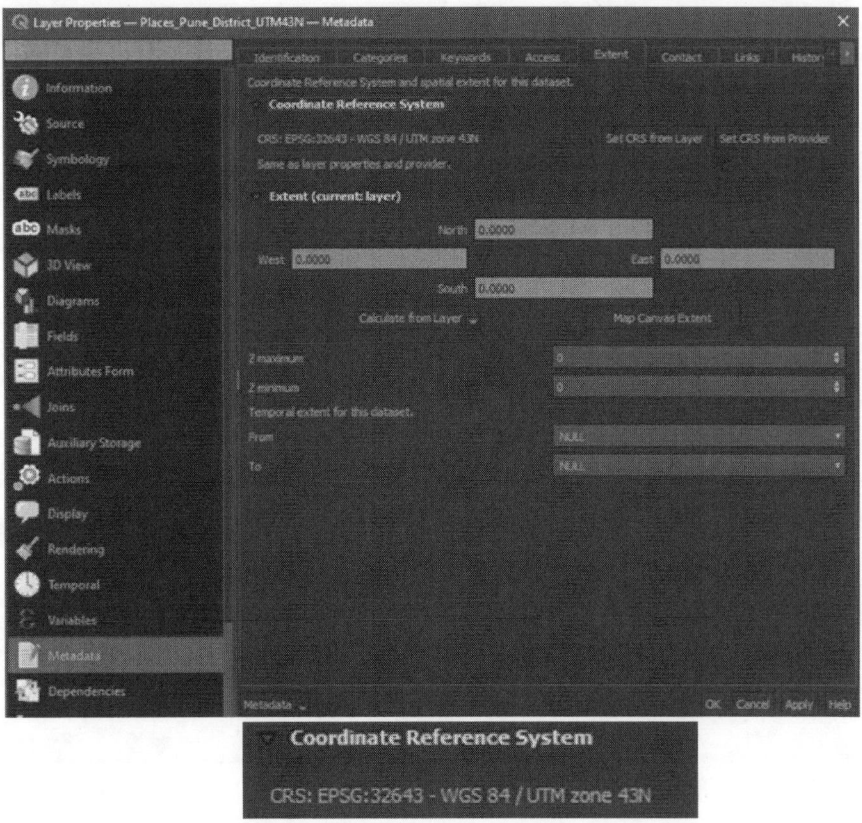

Figure 7.11: Step 11, Changed coordinate reference system

Projecting a raster dataset

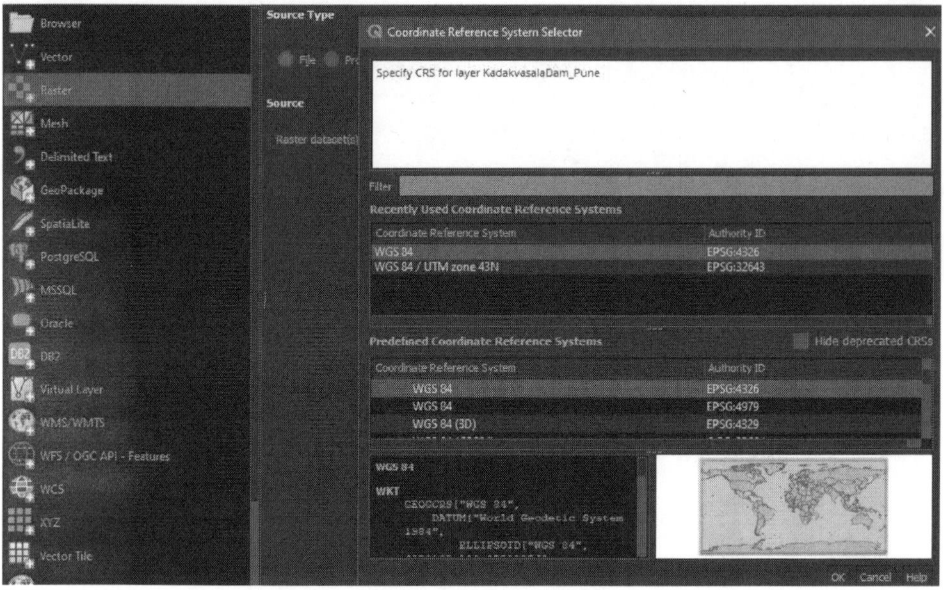

Figure 7.12: Step 14, Select the CRS WGS 84 datum

Step 12: If the raster data set isn't associated with any CRS then you will be prompted to choose the CRS'like Step 3.

Step 13: Add the raster layer of Kadakvasala Dam via, _Main menu bar → Data Manager → Add Raster Layer'or else directly click on icon from the toolbar

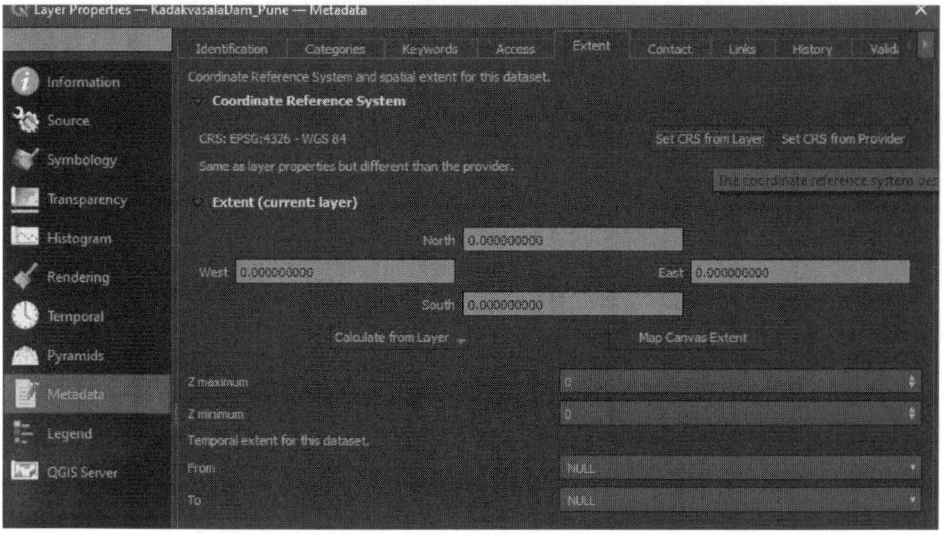

Figure 7.13: Projection of the raster data has been done

→ browse and select the raster layer named _KadakvasalaDam_Pune.tif' in tutorial data → Click on _Open' in the popup window.

Step 14: Since the _Kadakvasala Dam_Pune.tif' image doesn't contain any CRS information → you will be prompt to select CRS → Use filter to find _WGS 84' datum, select it → click 'OK'

Step 15: Now you can notice the _KadakvasalaDam_Pune.tif' layer is added to the _Map canvas' and listed under _Map Layers' section.

Step 16: Check the Metadata of the raster layer via, _Right-click on the layer Properties'. Click over _Metadata' tab in popup window → Select Extent then _Set CRS from Layer' to see the CRS details associated with it. You can see the layer has assigned to _WGS 84' datum.

Reprojecting a raster dataset

In this section we will reproject a raster layer from Geographic Coordinate System (GCS) toProjected Coordinate System (UTM).

Step 17: Open the georeference raster layer Geo_KadakvasalaDam_Pune.tif' as per Step 13.

Step 18: Now check for Layer SRS as per Step 16. It is in WGS 84 GCS projection.

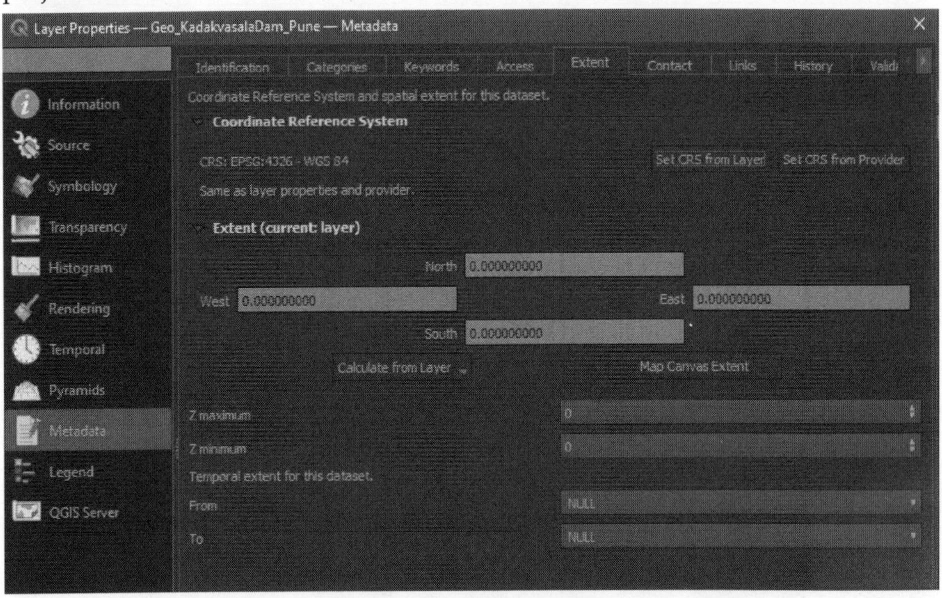

Figure 7.14: Step 18, check the CRS system of the raster data

Step 19: Go to the Main menu bar → Raster → Projections → Warp (Reproject)'.

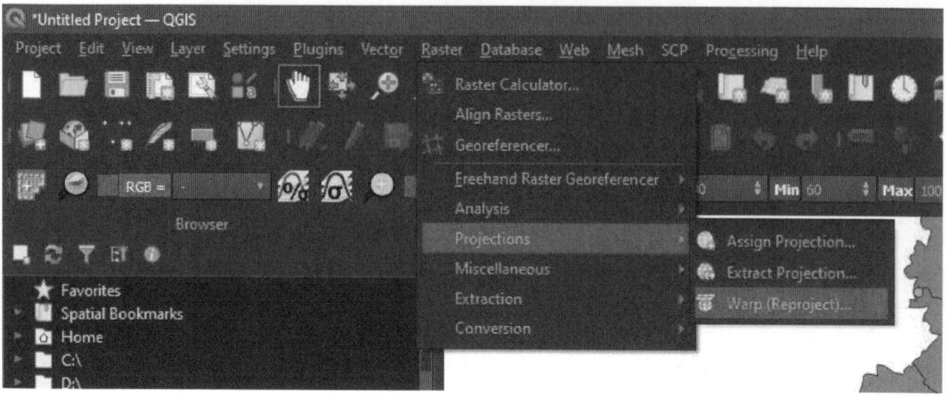

Figure 7.15: Step 19, Navigation to wrap (Reprojection) tool to reprojection of the raster data

Step 20: In the popup window of _Warp (Reproject)' → Select Input file as Geo_KadakvasalaDam_Pune.tif' → Go to Advance Parameters → click on Reprojected then _Save to File' _Select' of Output file → Browse to the desired path in popup window enter name as _UTM_Kadakvasala_Pune' and select Files of Type as → _[GDAL] GeoTiff (*.tif, *.tiff, *.TIF, *.TIFF)' click on Save'.

Now _Check' the check box of _Target CRS' → click on _Select' located right side to TargetCRS → use Filter to find _WGS 84/ UTM zone 43N' → select it click on _OK'

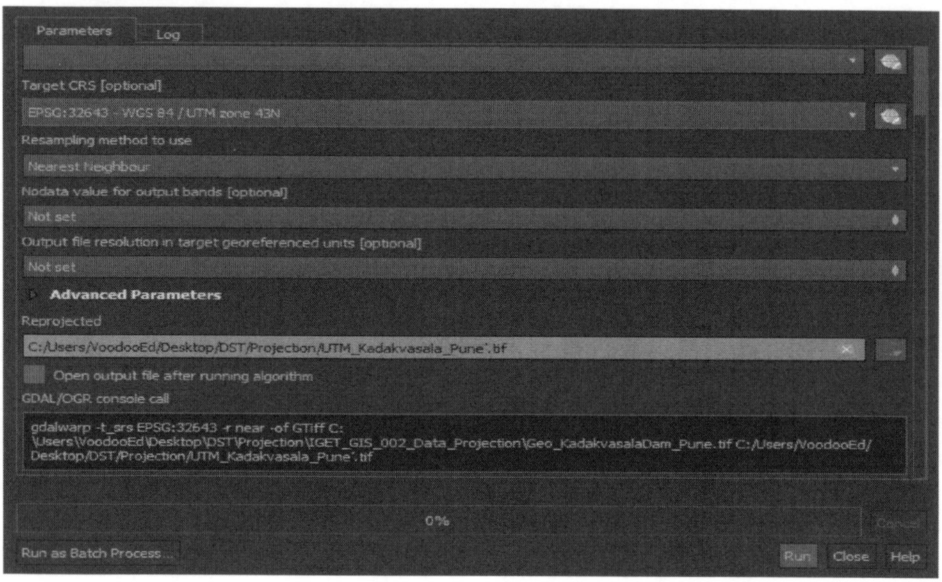

Figure 7.16: Step 20, Reprojection of the raster data

Step 21: Click on Run' in Warp window → you can see the _UTM_ Kadakvasala_Pune.tif'attached to the Map canvas.

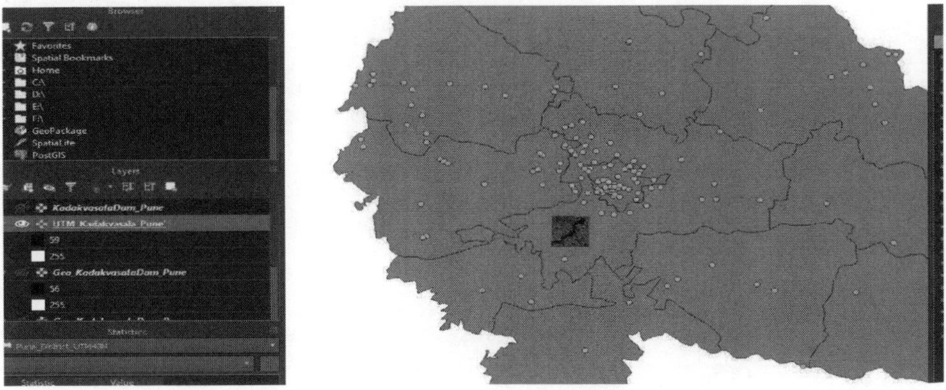

Figure 7.17: Step 21, Appearance of .tif file on the canvas

Georeferencing a Toposheet

What you should be able to do after this exercise?

We will get to know how to georeferenced a toposheet by using graticule intersections in aknown coordinate system and datum.

Principle

Georeferencing is a process of establishing a mathematical relationship between the image coordinate system and the real-world spatial coordinate system. This mathematical relationship can be assigned by any one of the transformation settings, viz. Polynomial order 1, 2 or 3, Linear, Projective and Thin Plate Spline etc. Polynomial order 2 is the most widely used transformation in Georeferencing. Recently Thin plate Spline gaining popularity due itsability of incorporating the local deformations in the data, this is very useful when we are dealing with low resolution data. However, in this practical we are using traditional polynomial order 2 transformation to perform georeferencing/rectification of the Toposheet. In this exercise, we will use a topographic Map of South Pune, prepared by Institute of Environment Education and Research (IEER), Bharati Vidyapeeth University (BVU), Pune. This map is in the Universal Transverse Mercator (UTM) projection based on WGS 84 Datum.

Procedure

Step 1: First, we open Quantum GIS (QGIS) via the Start menu. (Start → All Programs → QGIS Dufour → QGIS Desktop)

Step 2: We then open QGIS Georeferencer via the menu bar. (Raster → Georeferencer → Georeferencer).

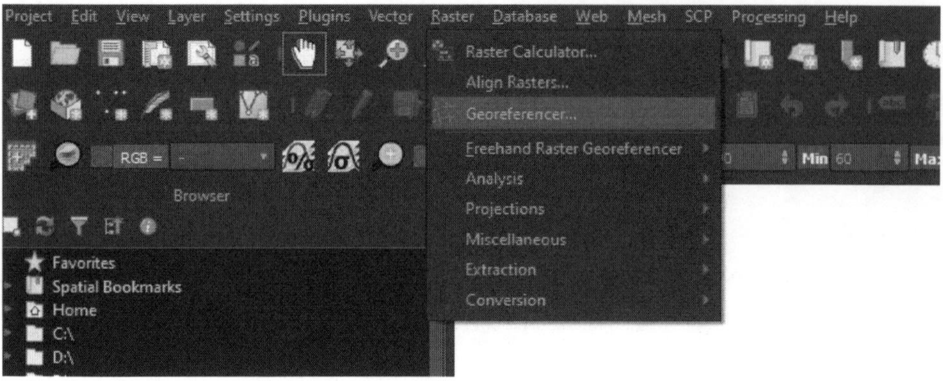

Figure 7.18: Navigating to georeferencer

Step 3: The Georeferencer is in the form of a window with two parts to it. The upper part is called as 'Main Workspace' dedicated to display the Raster Map to be georeferenced and it allows the user to input the geographic or projected coordinates of control points. The lower part, titled 'GCP table', is where the Ground Control Point data and residuals will be displayed.

Figure 7.19: Georeferencer window

Step 4: Add the toposheet to the Georeferencer by clicking on the 'Open Raster' button orfrom the File menu (File → Open Raster) or press ctrl + o.

Step 5: You will then be presented with a popup window, navigate to the tutorial data folder

in which the 'toposheet.tif' file is kept. Click on the drop-down menu right to the 'File name'and select '[GDAL] GeoTIFF'. Now select the _toposheet.tif' and click _Open'.

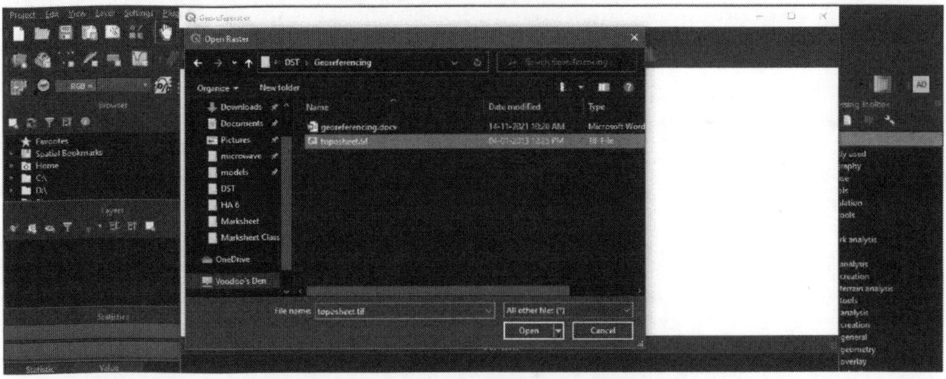

Figure 7.20: Add toposheet to georeferencer

Step 6: In the Georeferencer window go to Settings → Transformation Settings it will then be presented with the Transformation Settings ' window, from which we will select the 'WGS 84/ UTM Zone 43N' under Target SRS option.

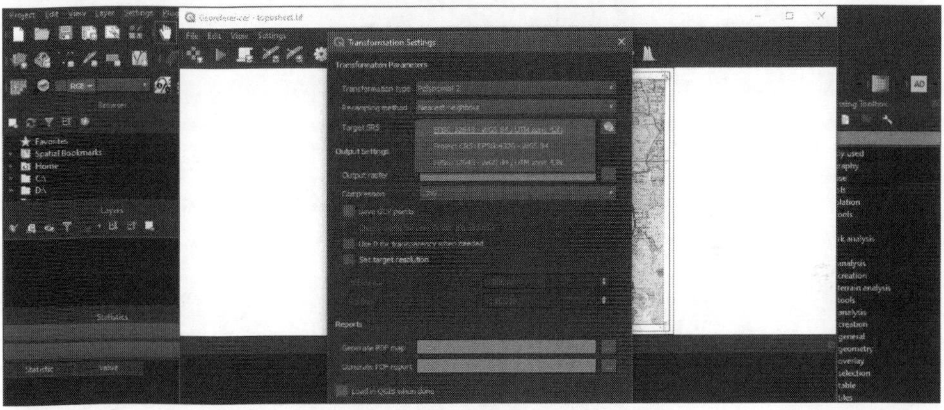

Figure 7.21: Transformation setting window

Step 7: The toposheet will be loaded in the Georeferencer window and should look like below.

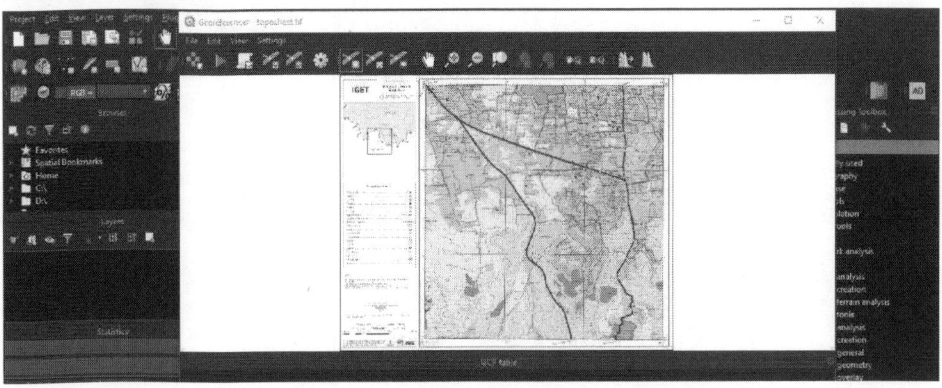

Figure 7.22: Toposheet in georeferencer window

Step 8: To georeference an image we use Ground Control Points (GCPs). GCP is a location on the earth's surface with known coordinates on both earth and Toposheet/ imagery, i.e., geographic and pixel coordinates resepctively. In this tutorial we use graticule intersections as GCPs.

Step 9: To start adding GCPs to our map, we first zoom to a corner of the map where we can easily identify the intersection of the latitude and longitude. Use the scroll wheel of the mouse to zoom in and out of the map. Use the 'Pan' Button when needed.

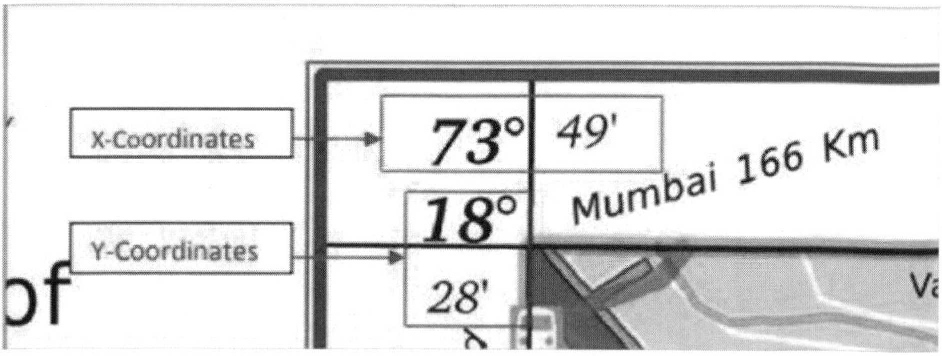

Figure 7.23: Zoomed in toposheet to add GCPs to the map

Step 10: To add a GCP click on the 'Add point' button, or go via the menu (Edit Add Point). The mouse will transform into a '+' sign, which we use to click on the centre of the intersection. Use 'View tool' when needed.

Figure 7.24: Step 10, Add GCP

Step 11: A window 'Enter map coordinates' will pop up where we enter the coordinates of the point which we take from the map. Always enter 'Longitude or Easting' in X field and 'Latitude or Northing' in Y field. Use 'space bar' in the key board to separate the Degree value from Minutes value and then click _OK'.

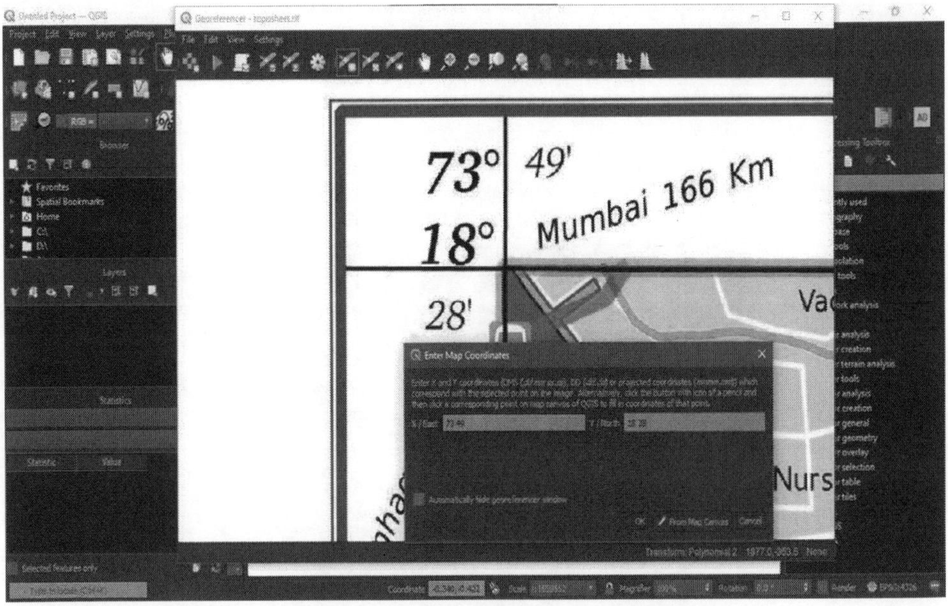

Figure 7.25: Step 11, Entre map coordinates

Step 12: In this exercise we are using the 'Polynomial 2' transformation to georeference the image. For the 'Polynomial 2' transformation we will require minimum 6+1(for check) i.e., 7

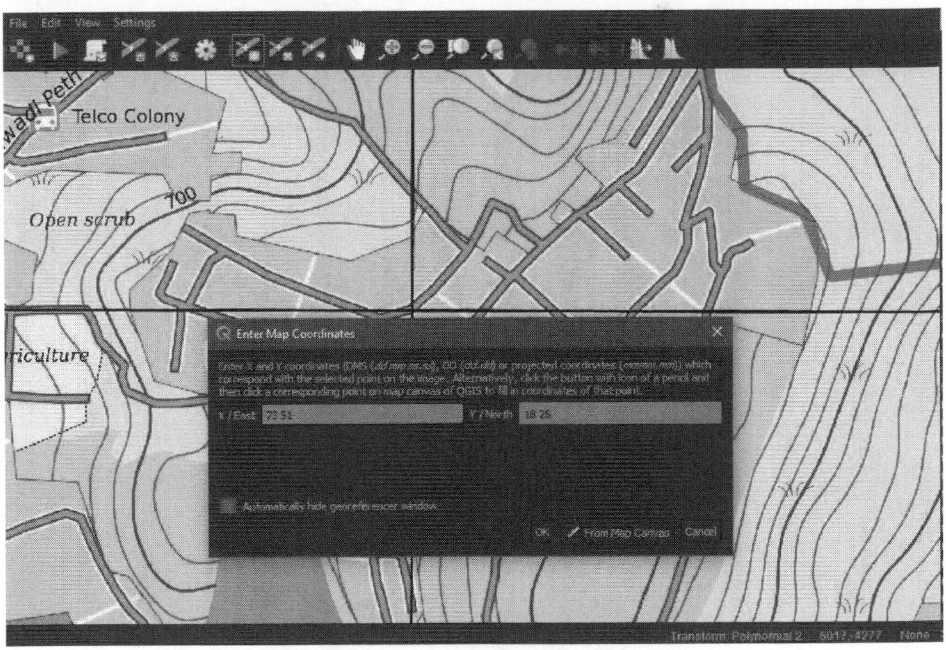

Figure 7.26: Multiple GCPs

GCPs or more GCPs on the map. Therefore, we need to mark at least 7 GCPs. The GCPs locations should be spread out as much as possible and they should not be co-linear at the same time they should enclose our whole area. Use the above procedure to mark six more control points.

Step 13: We now set the spatial reference settings for the toposheet by clicking on the 'Transformation Settings' button. The 'Transformation Settings' window pops up in which wewill enter the spatial information of our map.

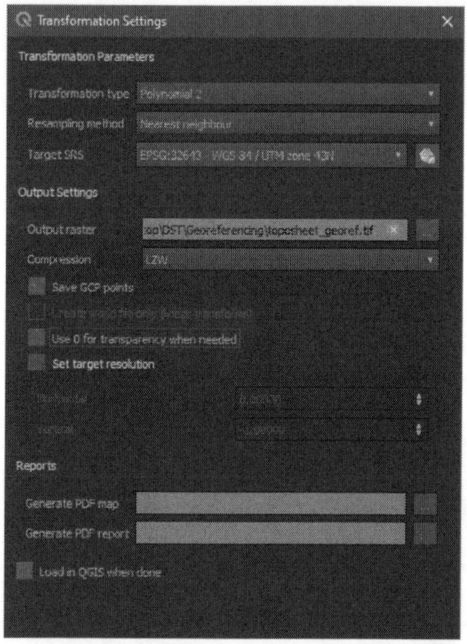

Figure 7.27: Enter spatial information in the transformation settings

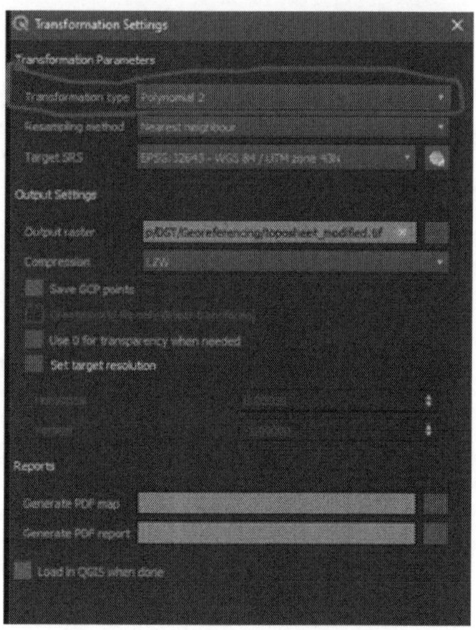

Figure 7.28: Second order of polynomial transformation

Step 14: Click on the 'Transformation type' drop-down menu and select 'Polynomial 2'. This means we will be using a second order polynomial transformation.

Step 15: Click on the button next to Output Raster'. A dialogue box will appear in which weenter the name of our output file. It is recommended to include the name of the original file in this file for example 'Toposheet_WGS84_georef.tif ', this helps us to keep track of our work.

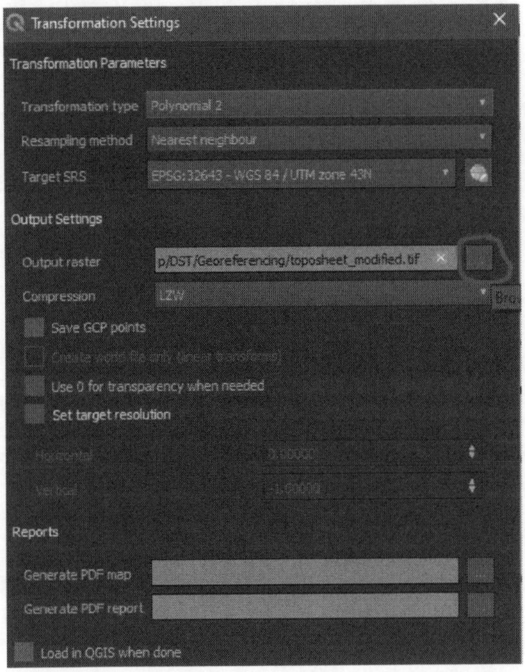

Figure 7.29: Rename and save the output at the selected location

Step 16: Check' the check box 'Load in QGIS when done' and 'Use 0 for transparency whenneeded', leave the reset of the values as default and click 'OK'

Step 17: After Click 'OK', the last column 'residual [pixels]' displays some values. These are the error values associated with the GCPs. An error value of 1 or less would be satisfactory.

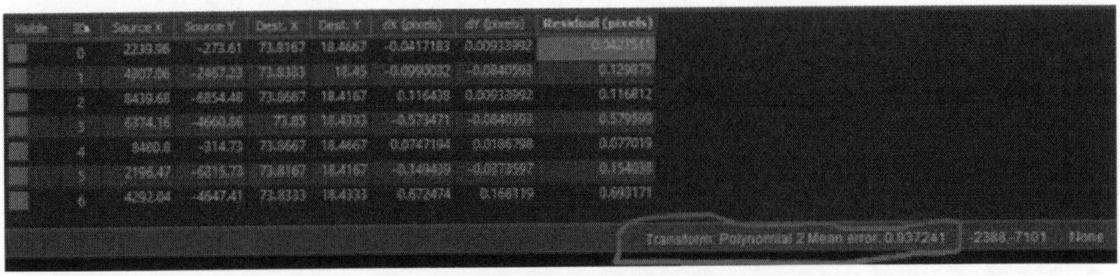

Figure 7.30: Error value table

Step 18: Double check and adjust the GCP locations if the value is greater than 1. To adjust a point click on the 'Move GCP Point' button, and then click and drag the point to the desired location. Use 'Delete Point' button to delete an erroneous GCP. We can also enable and disable the GCPs by using check boxes under 'on/ off' column in 'GCP Table'.

Step 19: Once the error is around or below 1, click the 'Start Georeferencing' Button. The processing will take at about 2 minutes. The georeferenced image will be found at the location specified for the 'Output File' in step 15. You can also notice, the output file loaded in QGIS canvas.

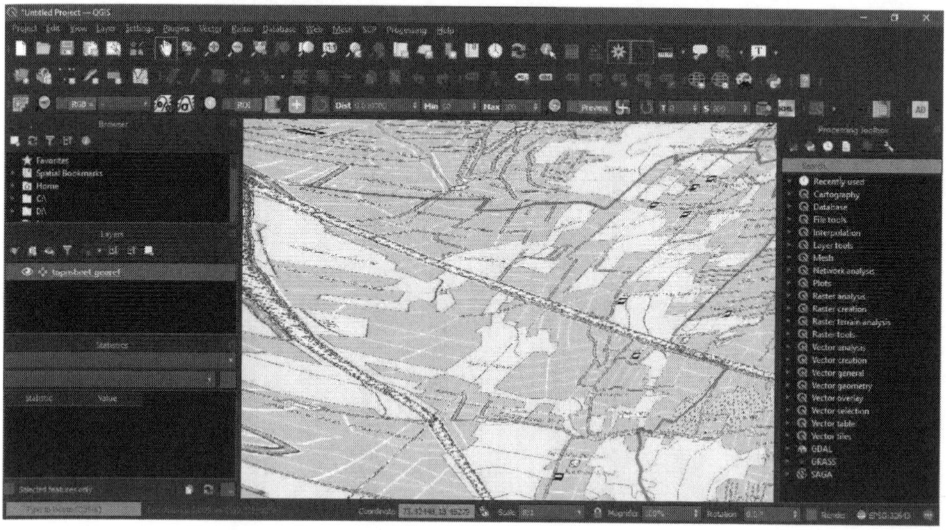

Figure 7.31: Georeferenced toposheet

Step 20: It is useful to save the GCPs for later use if the georeferencing needs to be done again, or if corrections are required. To save them, click on the 'Save GCPs as' button. In the pop up window enter an appropriate name for the GCP file, preferably the same as that of the image. This file can be loaded later on by clicking on the 'Load GCP points' button in 'Georeferencer' window after adding toposheet, and selecting it from the pop up window.

Map Preparation using QGIS

What you should be able to do after this exercise?

We will be able to know how to make a map and how to add legends, north arrow, and textin a map and how to convert a map into a pdf file.

Principle

Map is a graphical representation of a particular area or whole earth on a flat surface. This flat surface may be a paper or a cloth etc. A very common output of any GIS/Surveying work is map. A map conveys more information than words, therefore a good map portrait your results and efforts in a better position. Cartography is a branch of art and science, whichdeals with the map making. There are a wide variety of maps available topographical, geological, thematic, climatic, military and guide maps are few among many. The main objective of this tutorial is to guide the user how to prepare a basic map in QGIS. In this tutorial we will learn how to label and edit the symbology of various features to create a mapshowing hospitals and bus stops of south Pune.

Procedure

Step 1: Start QGIS and open the map layers via the menu Layer → Add Vector Layer. In the next window, click on _Browse'. Navigate to the tutorial data folder with the vector layers inthe popup window. Change the layer type to _ESRI Shapefiles'. Select all the layers by clicking on one and pressing CTRL+A. Click _Open' for both windows.

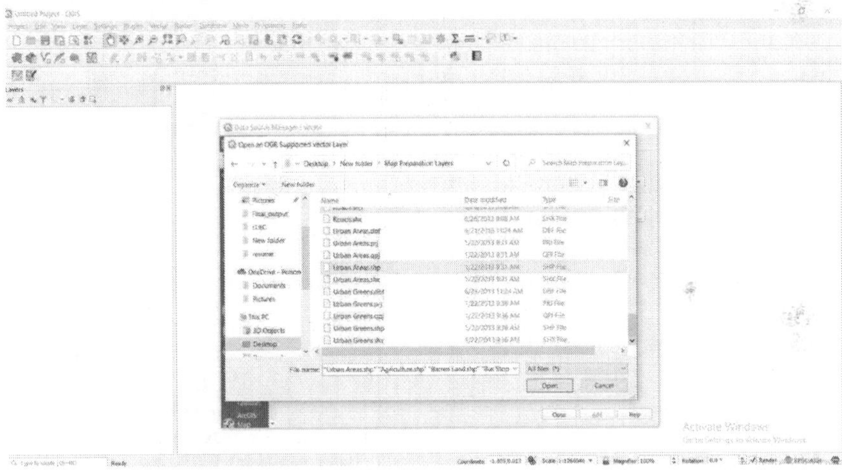

Figure 7.32: Adding vector layer to the canvas

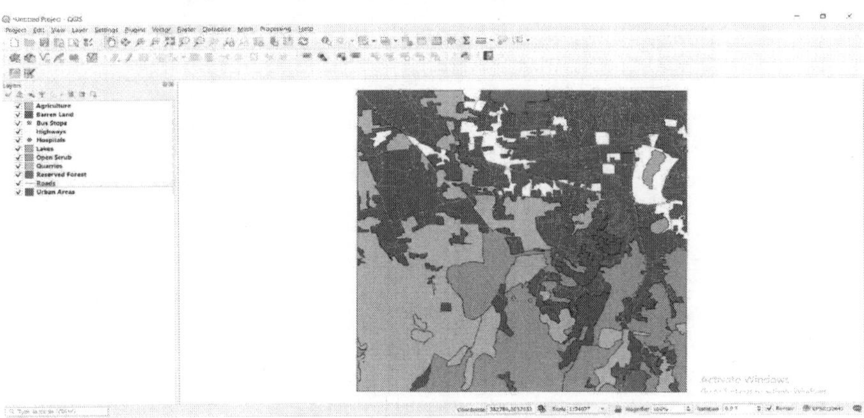

Figure 7.33: Added vector layers

Step 2: Now all the layers will be loaded to the map canvas and will be listed under the Layers section alphabetically in descend order. This is because QGIS loads the files the order in which they are stored on the computer.

Step 3: Now rearrange the layers in such a way, all point features are on top to line features and line features over the polygon features. This can be done by selecting the layer and hold the left mouse button over it (the mouse cursor will turn into to a double headed arrow) and then dragging them up or down.

Step 4: Let's start the map preparation by setting the colors for the base polygon layers. Double-click on the Urban Areas' layer to open the Layer Properties' window and select the –Symbology/Style‖ tab.

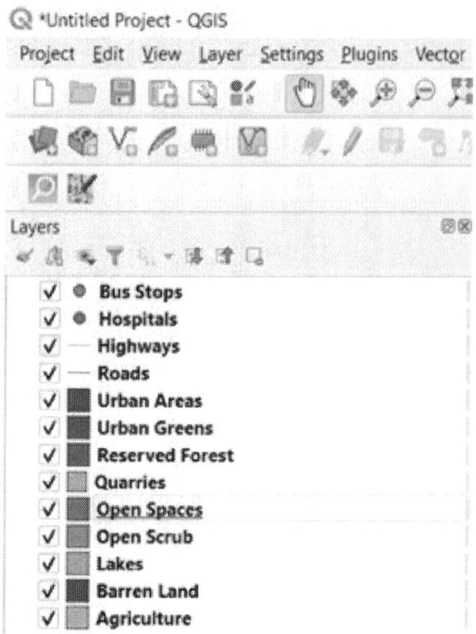

Figure 7.34: Rearrangement of the layers

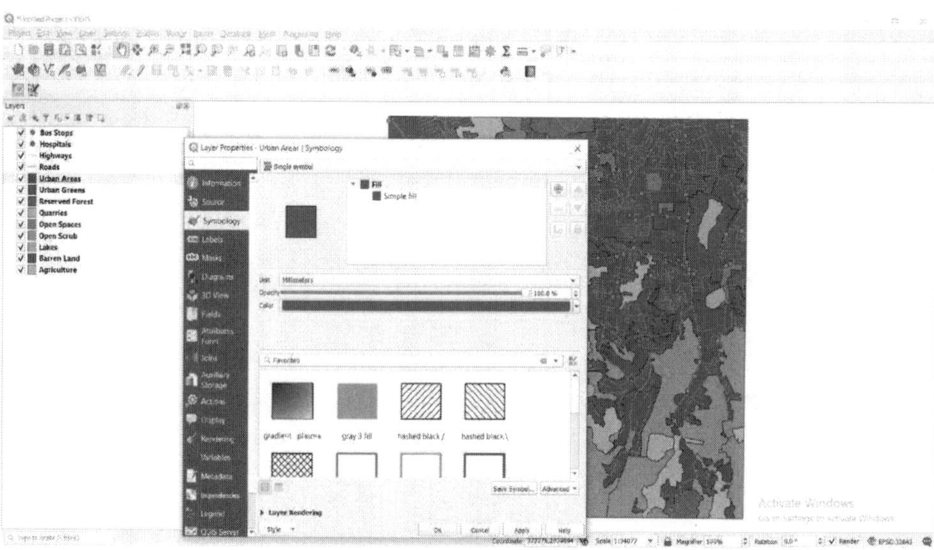

Figure 7.35: Change the color of the polygon layer

Step 5: While setting the colors for polygons, it must be remembered that they are larger thanmost other features in the map and their presence is already felt by virtue of their size. We must therefore assign light and pastel colors to make sure the polygons fall into the background.

Figure 7.36: Selection of colors

Step 6: Now you will notice the Urban Areas color turned to light shade of pink.

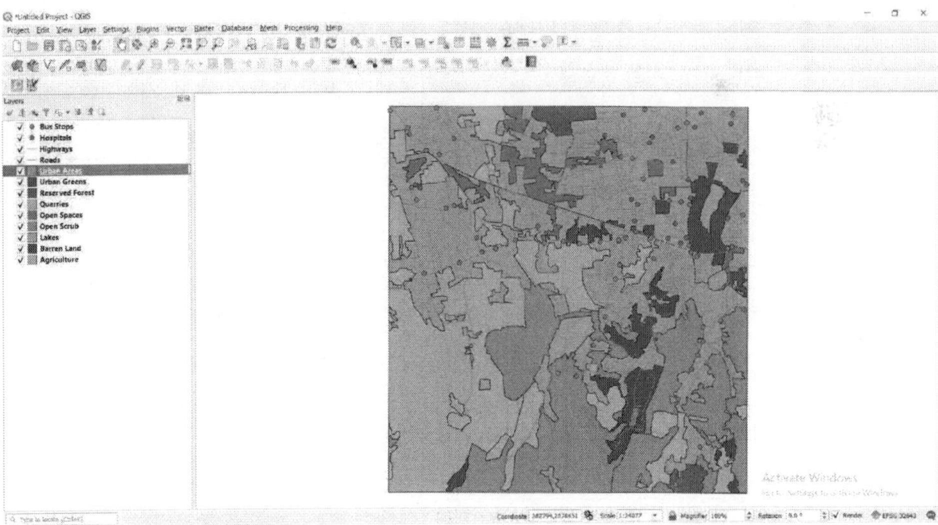

Figure 7.37: Color changed

Step 7: Use the above same method to set the colors for all other the classes.

Step 8: Now the map should look like near to the below figure, however it depends upon your assorted color code.

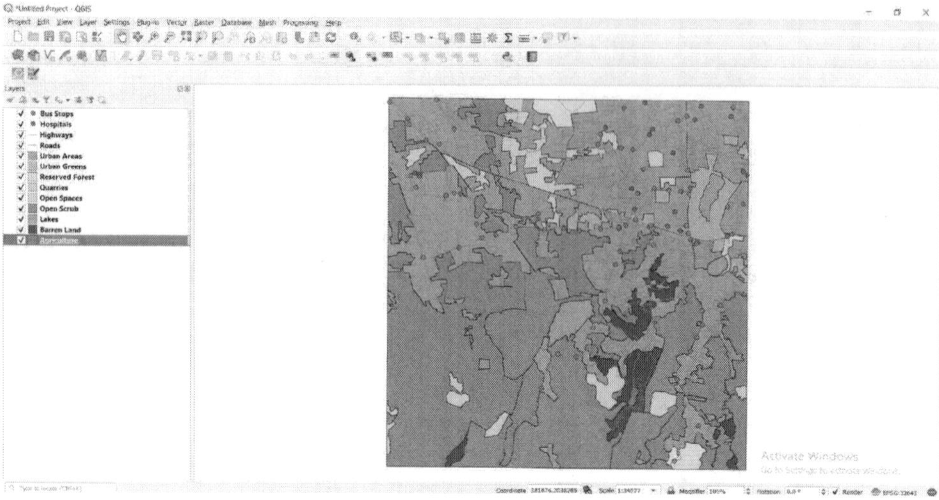

Figure 7.38: New map after color change

Step 9: Take a look at the _Open Scrub' and _Urban Greens' layers. They both have the somewhat similar color. We can differentiate them by simply changing the colors, sometimes patterning yields a better visualization then simply changing colors. Open the _Layer Properties' window of the _Urban Greens' layer. Click on the _Add Symbol Layer' button on the bottom left of the window. This will add a new symbol layer under _Symbol layer' section. Select it and change the _Symbol layer type' to _SVG Fill'. Use the symbol from the SVG symbols, change the color and texture width as shown in the above figure

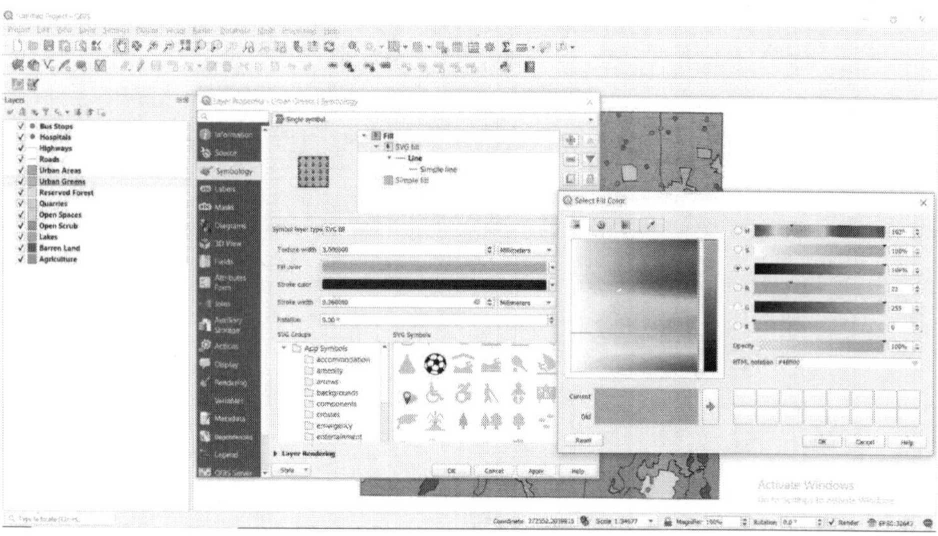

Figure 7.39: Change the pattern of the polygone for better visualization

for the _Urban Green' layer and click _OK'. Use the same method to set your desired symbologyto other polygon layers.

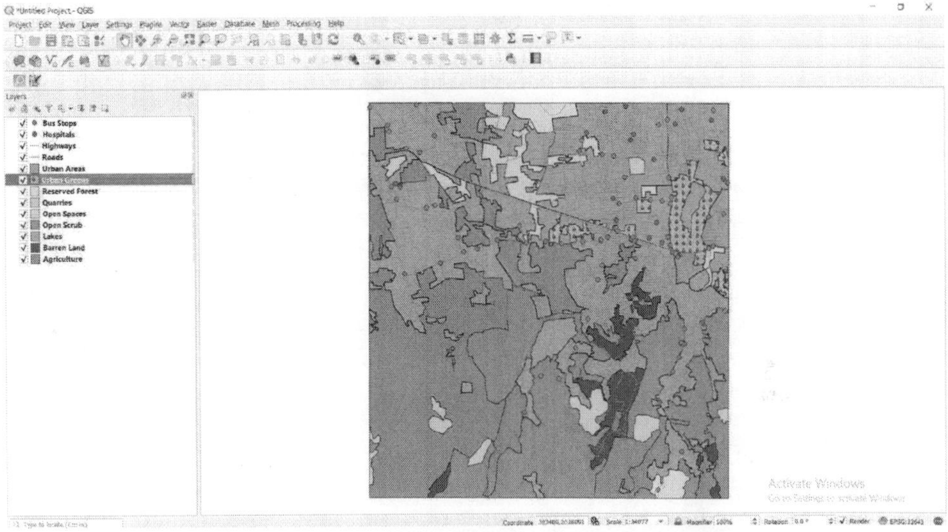

Figure 7.40: Changed pattern of the polygone

Now that we've set the symbology of the polygons, we will start with the lines. For the roads we will use the colour orange with a red border.

Step 10: Double-click on the Roads' layer to bring up the Layer Properties' window. To emulate the effect of having an orange fill with a red boundary, we will use two symbol layers. The thinner orange layer will be stacked on top of a thicker red layer.

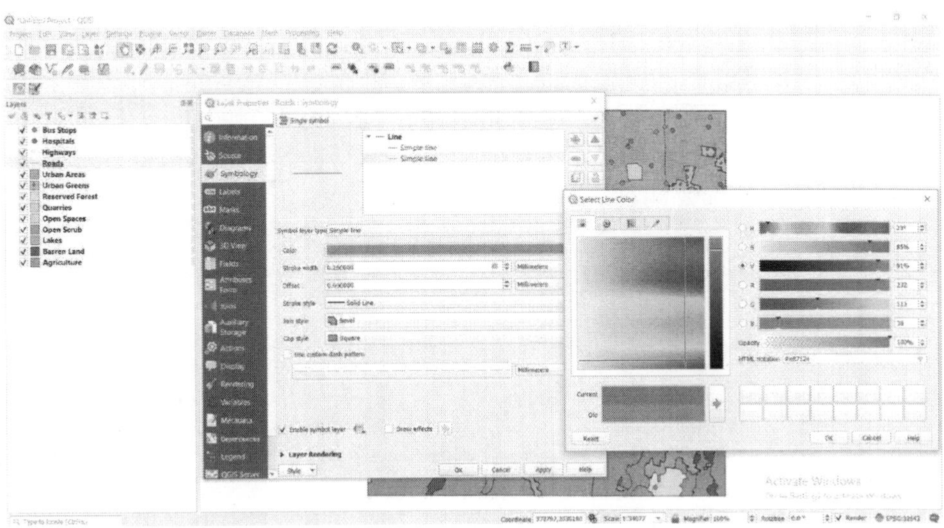

Figure 7.41: Change the color of line

Use the _Add Symbol Layer' button to add another symbol layer. Select the upper layer andchange its color to _orange' and its width to _0.7mm'. Set the lower layer with a width of 1mm' and color _RED' and click on _OK'.

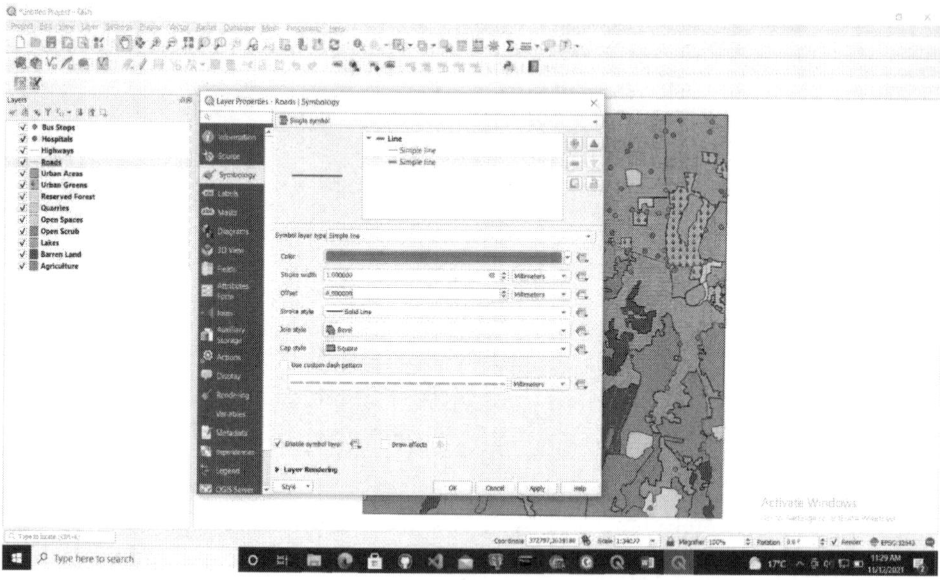

Figure 7.42: Change symbol for the feature

Step 11: For the _Highways layer' we will use the same approach. Change the layer symbol to a _red' line of width 3mm stacked on a black line with width 3.2mm.

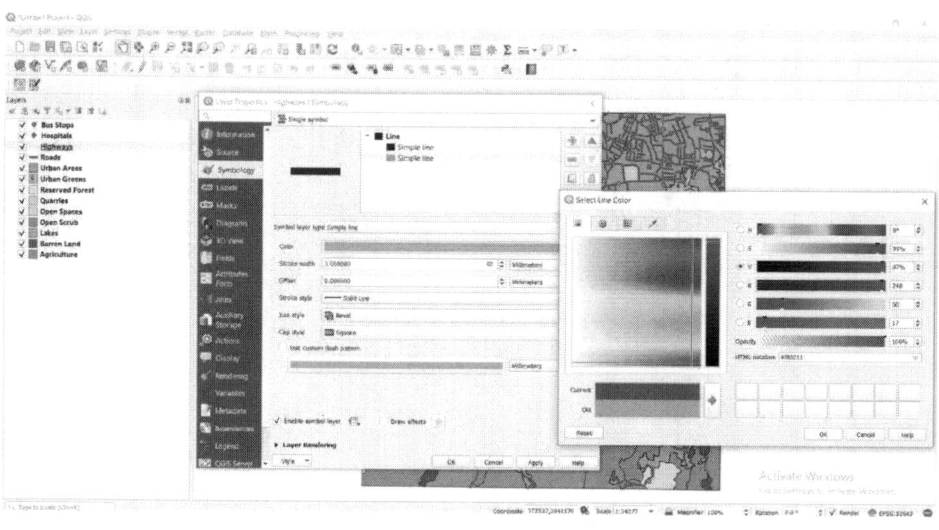

Figure 7.43: Change the symbol for the highway

Step 12: If you look at the road or highway junctions you will notice that the intersections are not contiguous. To smoothen the intersections, open the Layer Properties' window of Highway layer and from _Advance' the menu (locate just above the right-bottom of the window) select _Symbol level'. In the popup window, check _Enable symbol levels'. Click OK' on both windows. Do the same for Road layer.

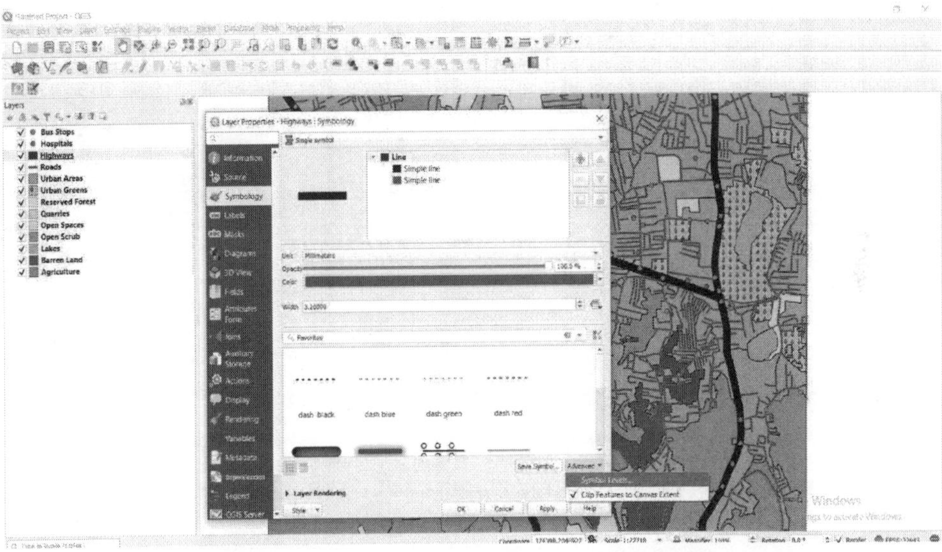

Figure 7.44: Smoothening of the intersection

Step 13: The point layers are currently symbolized by colored circles. Although they are easily identified when the circles replaced with more pictorial/descriptive

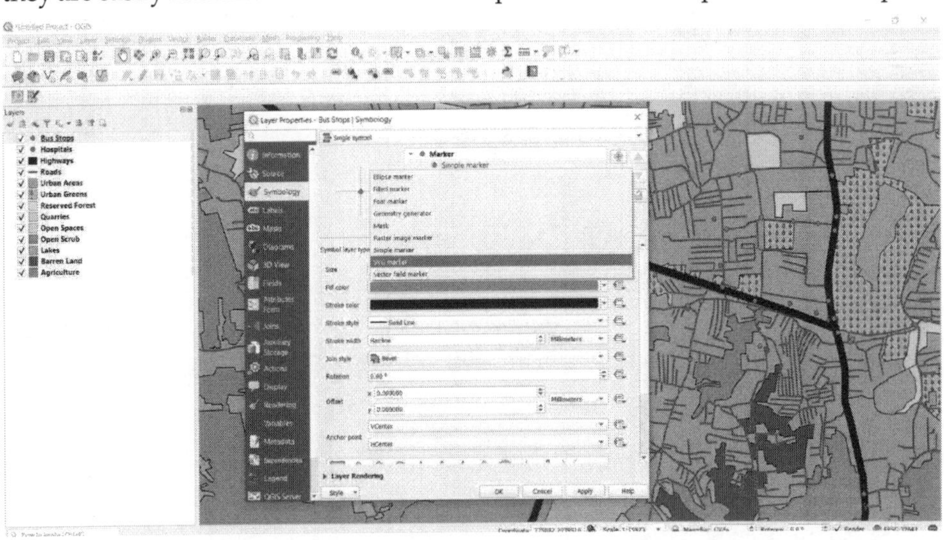

Figure 7.45: Change the symbols

symbols. Double click the Bus stop' layer and select the _Simpler Marker' under _Symbol layers'. Change the Symbol layer type' to _SVG Marker' and use the bus stop icon and change the other fieldsas shown below and click on _OK'.

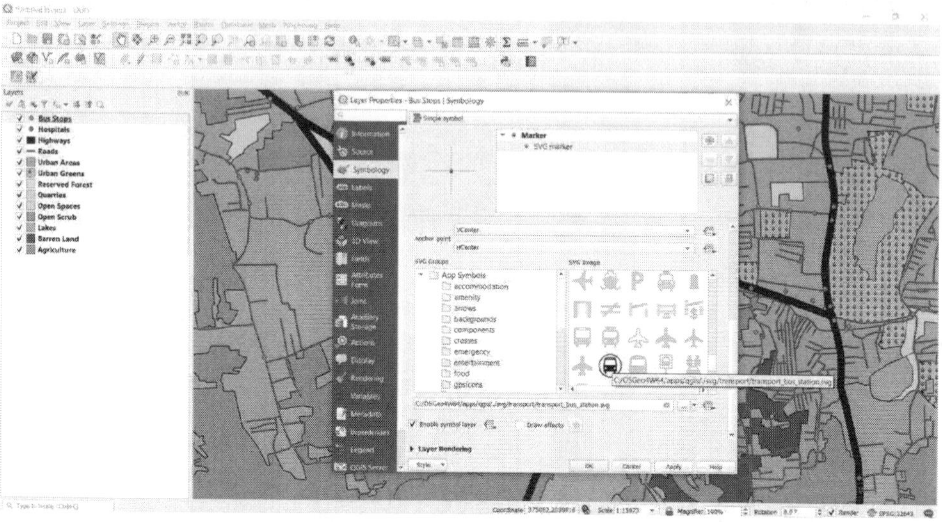

Figure 7.46: Selection of the symbol

Step 14: Follow the above same procedure to change the symbology of Hospitals' layer.

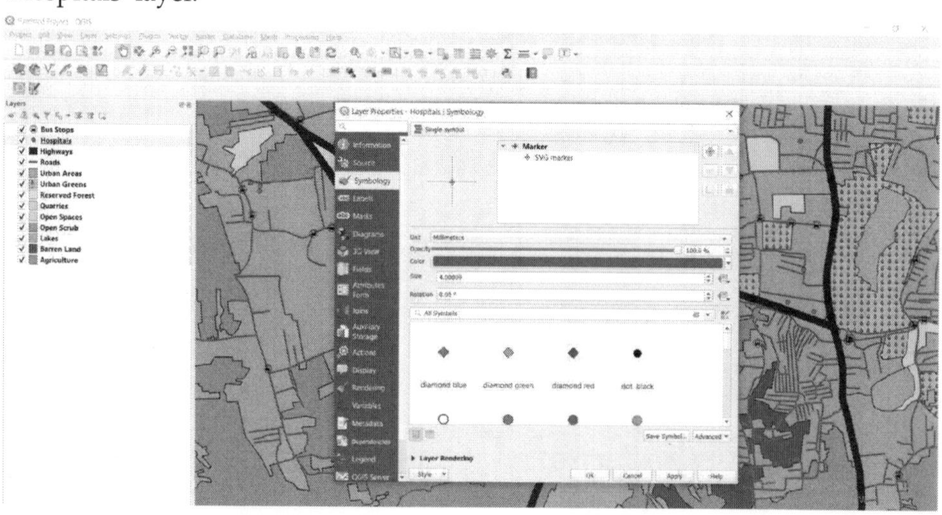

Figure 7.47: Change the symbol of hospital

Step 15: Click the Zoom Full' button to view the entire map. Our map should now look likebelow and ensure the layer order under Layer section as same as below.

Figure 7.48: Zoom in for better visualization

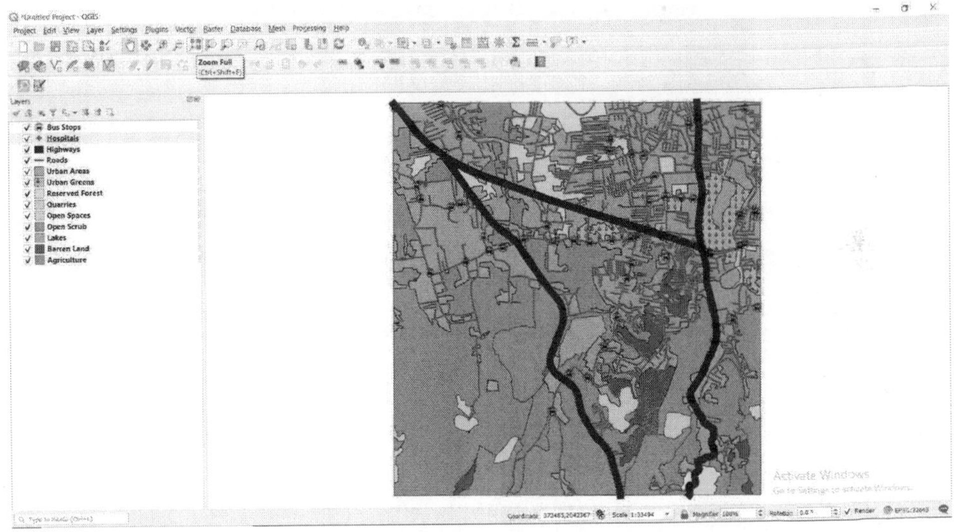

Figure 7.49: Full map after changing the symbols and colors of different features

Step 16: Now we label the point layers using the names provided in the attribute table. Right click on the Hospitals' layer and select Open Attribute table'. Here we see that the name ofthe hospital associated with each place mark in shape file.

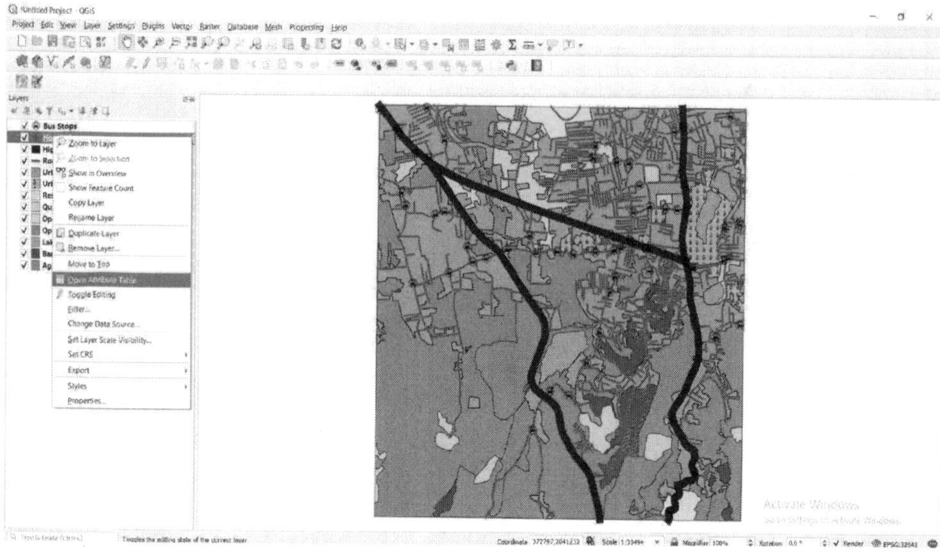

Figure 7.50: Labeling of the point features

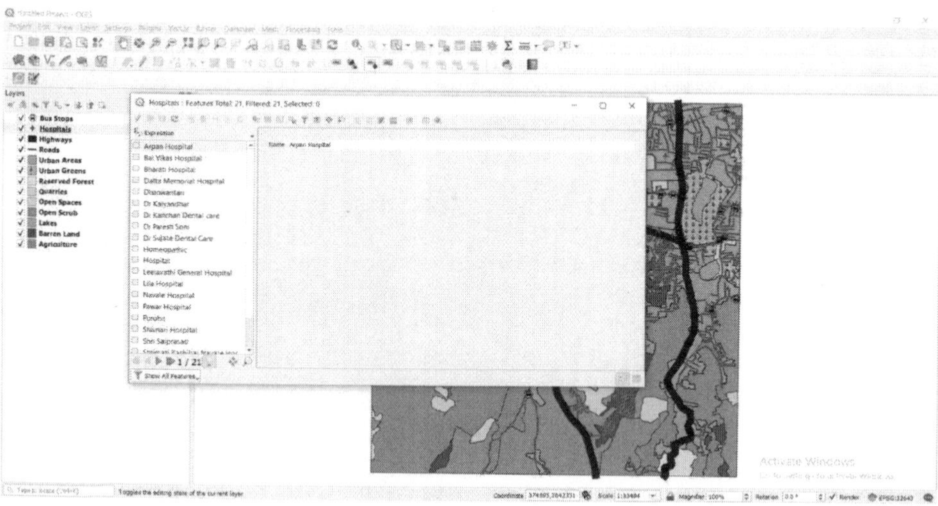

Figure 7.51: Labeling the hospital

Step 17: To label the features we will use the Label engine provided in the Label Toolbar. If the tool bar is not present activate it via the menu View → Toolbars → Label.

Step 18: Select Hospitals' layer under _Layers' section. In the _Layer label settings' popup window check the option _Label this layer with' and select _Name' from dropdown menu next to it and Click _OK'.

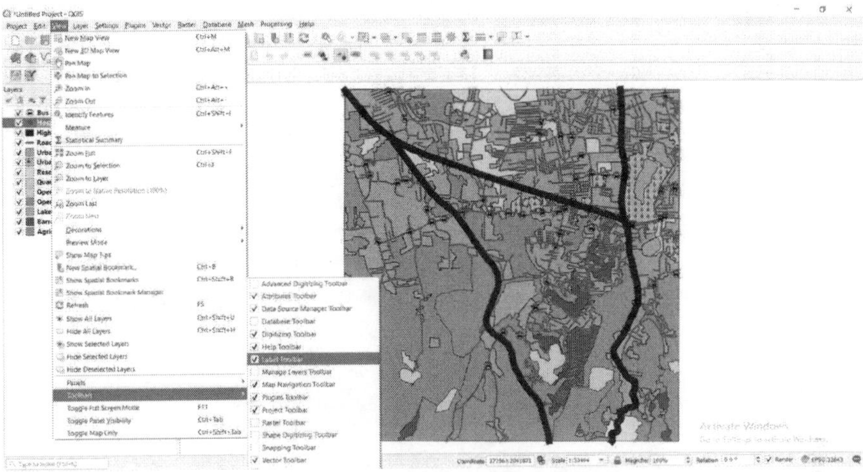

Figure 7.52: Enable the label toolbar

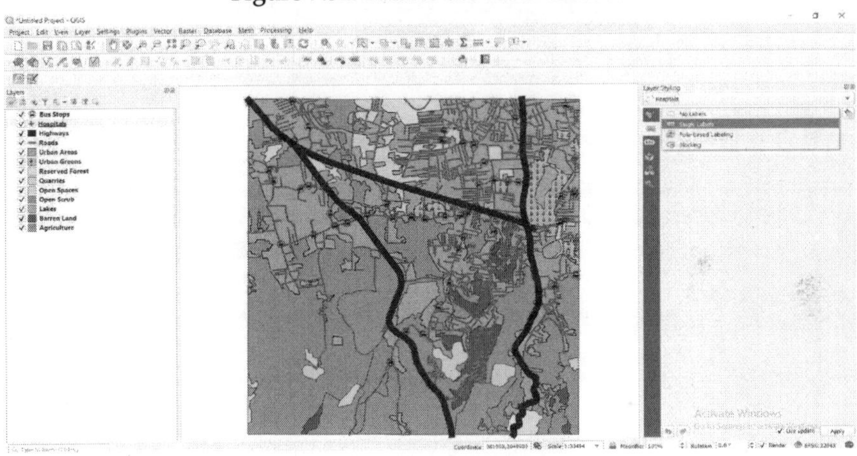

Figure 7.53: Select single layer

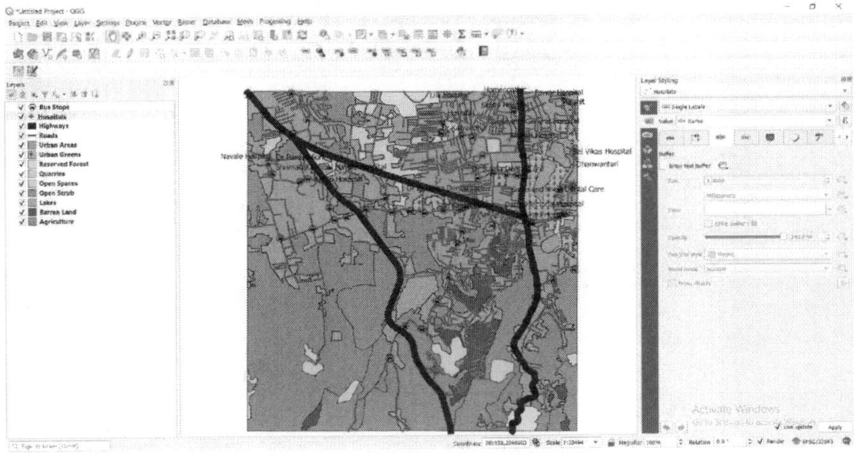

Figure 7.54: Hospitals with its labels

Step 19: Do the same for the Bus Stops' layer.

Step 20: This is how the labelled features will appear on the map:

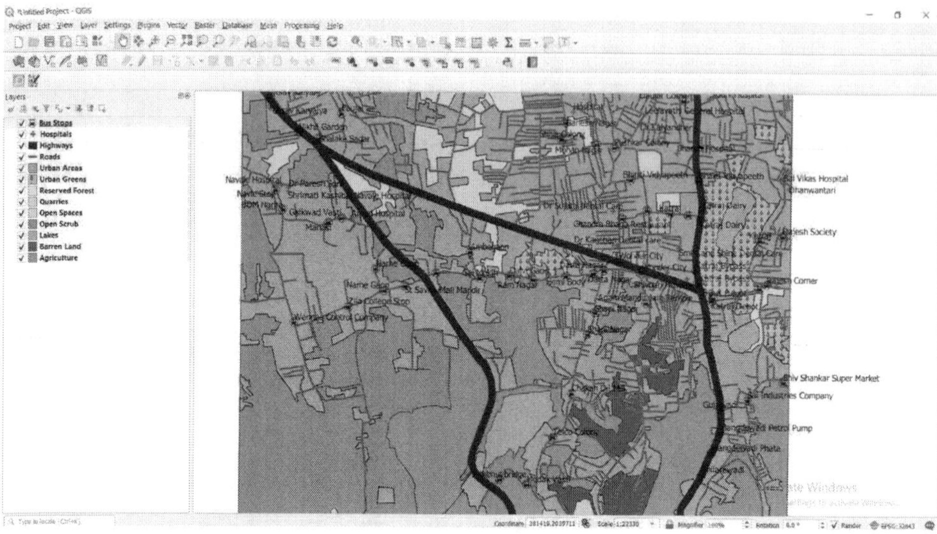

Figure 7.55: Labels of each layer

Step 21: To prepare a map use the Print Composer. Open it via the menu Project → New Print Layout/ Composer. You can assign a unique print composer title. Call it _SouthPune' and click on _OK'.

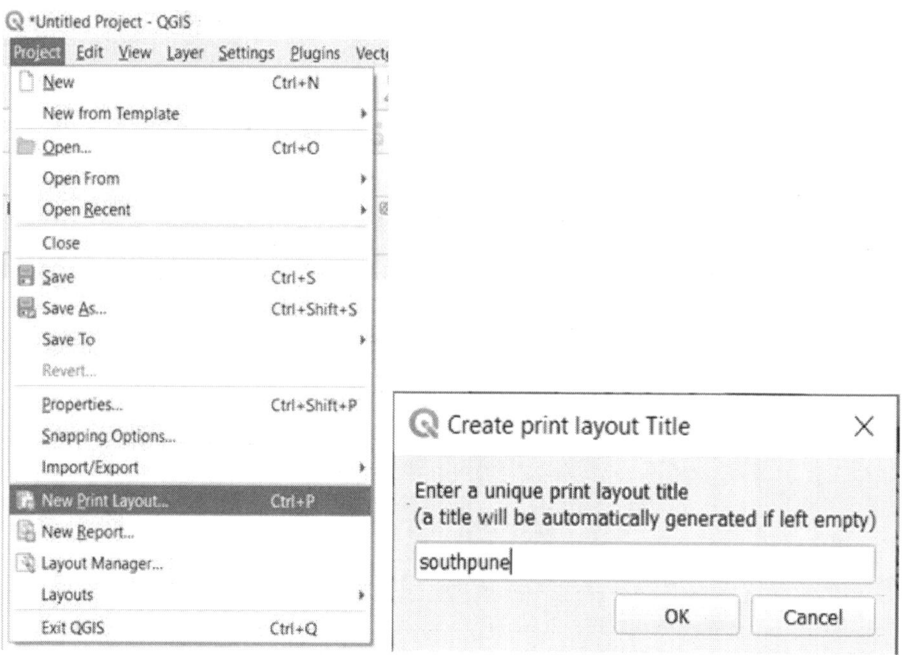

Figure 7.56: Print layout tool to print the map

Step 22: Change the size of the paper to A2 (420x594mm)' using the Composition tablocated right side inthe composition window and clickon Zoom Full' inthe tool bar.

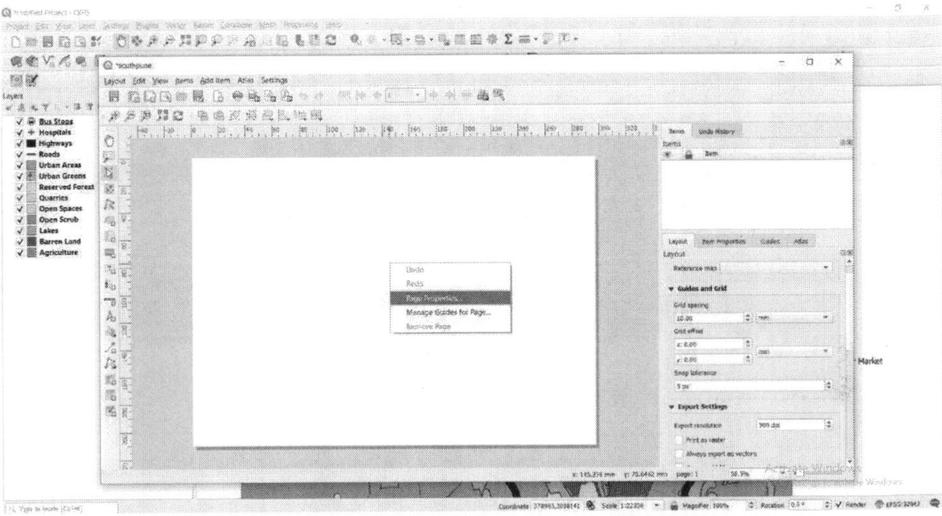

Figure 7.57: Composition tab to change the size

Figure 7.58: Page size

Step 23: Click on the Add new map' button in the toolbar and place the mouse pointer on the top-left corner of the composers' workspace and hold the left mouse button to drag across the whole page to create the map.

Figure 7.59: Add new map

Step 24: The map will be displayed in the composer's workspace generally it would be at a high or low scale than the publishing scale. In this tutorial 1: 15000 will be considered as publishing scale, to set this scale click on _Item Properties' tab and type _15000' in the Scale' field and press _Enter' in the key board.

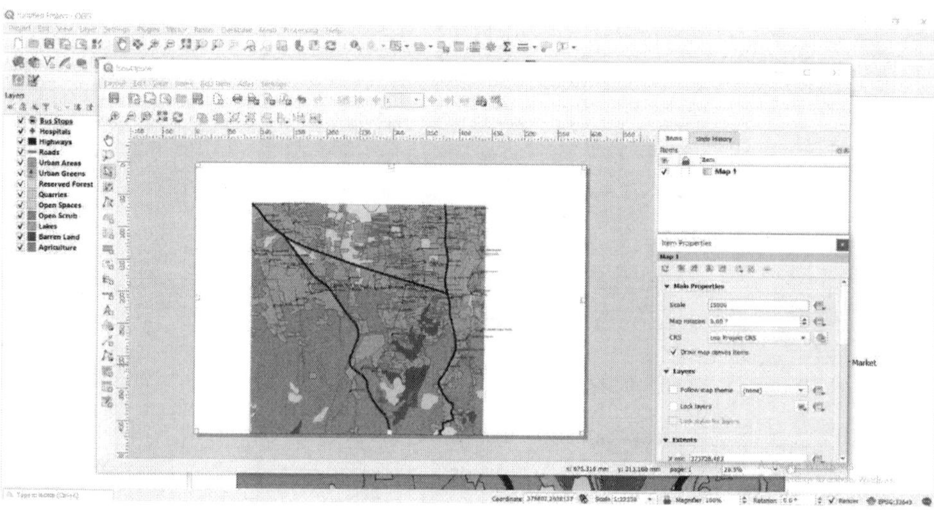

Figure 7.60: Map in composers workshop

Step 25: Click on the Move Item Content' button in the tool bar, and drag the map to a suitable position like shown below.

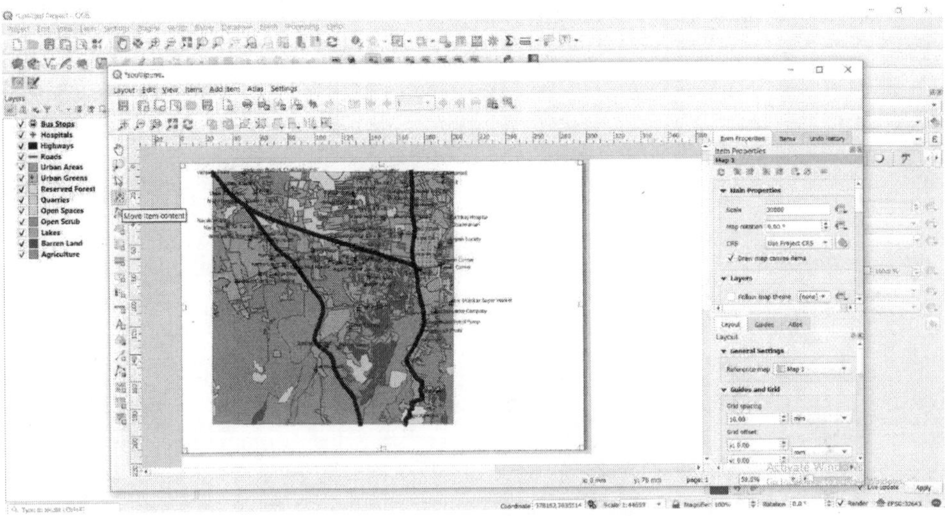

Figure 7.61: Dragging the map

Step 26: A map must contain Title, Author, Legend, Date, Orientation and Scale. These are the minimum necessary elements for any map. Add the title by clicking on the Add New Label‘ button and then clicking on a location. Once it is placed, click and drag the corners to increase the size. Click on the Label box in Item Properties‘ and in the entry, box enter the title of the map as Hospitals and Bus-Stops of South Pune‘. Click on the Font‘ button to change the font style and size to Bold‘ and 36 points‘ respectively. Set the horizontal and vertical alignment to Centre‘ and Middle‘ respectively. Now resize the rectangle box to fit the title text.

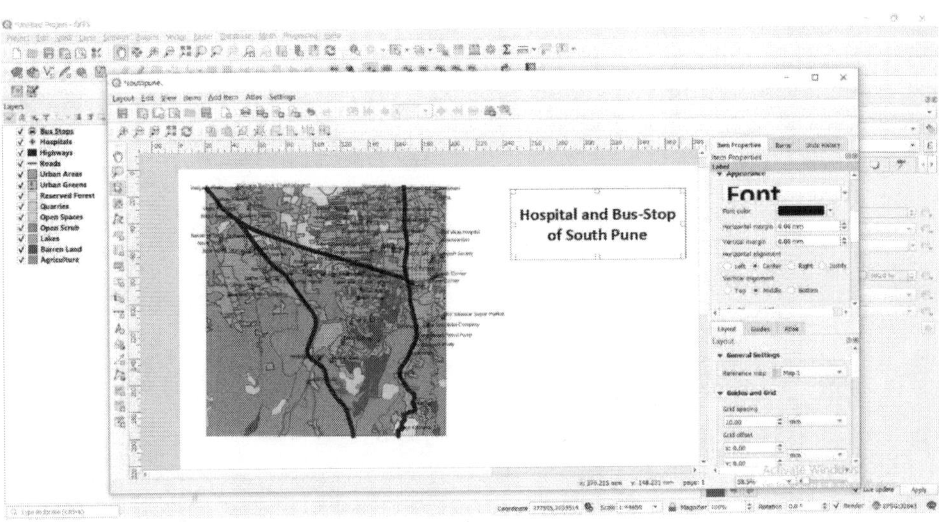

Figure 7.62: Add title to the map

Step 27: To add legend click on the Add new Legend' button in toolbar and place the mousecurser over a suitable place in the workspace and click on left mouse button to place legend.

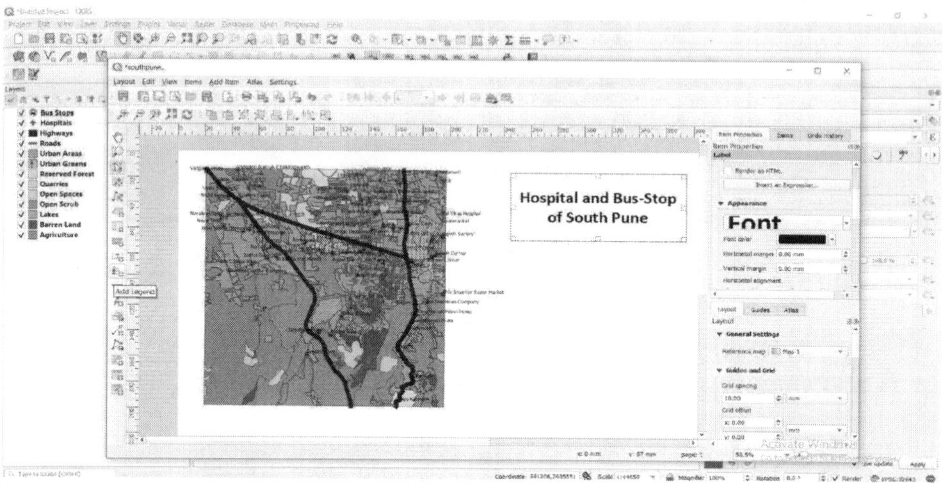

Figure 7.63: Add legend tool on the left corner

Step 28: To add the scale bar, click on _Add new scale bar' button and place it on the map. Set the Segment Size' to 750 and the _Map units per bar unit' to 1 and all other values as shown below figure. These are two important parameters:

a. _Segment Size' – This defines the length of the scale bar segment in map units (in our case it is meters).

b. Map units per Bar Unit – This translates the map units to the scale bar units. Since they both are meters, we will keep this as 1.

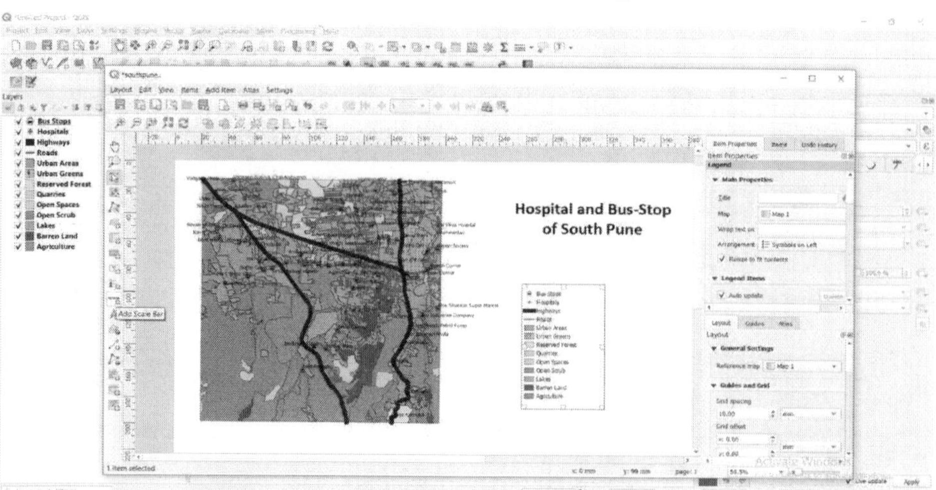

Figure 7.64: Add scale bar tool on the left corner

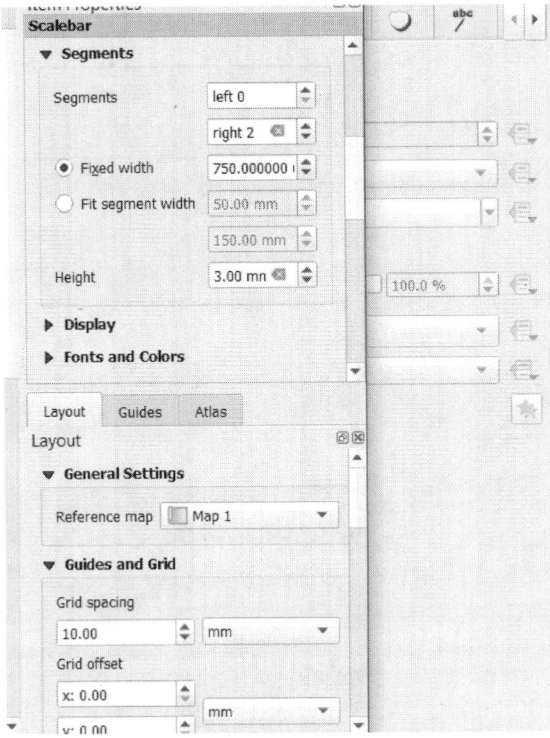

Figure 7.65: Settings of the scale bar

Step 29: For the north arrow, click on _Add image' button. Place it in the top right of the map. From the Item Properties', click on _Search Directories' to select one of the north arrows.

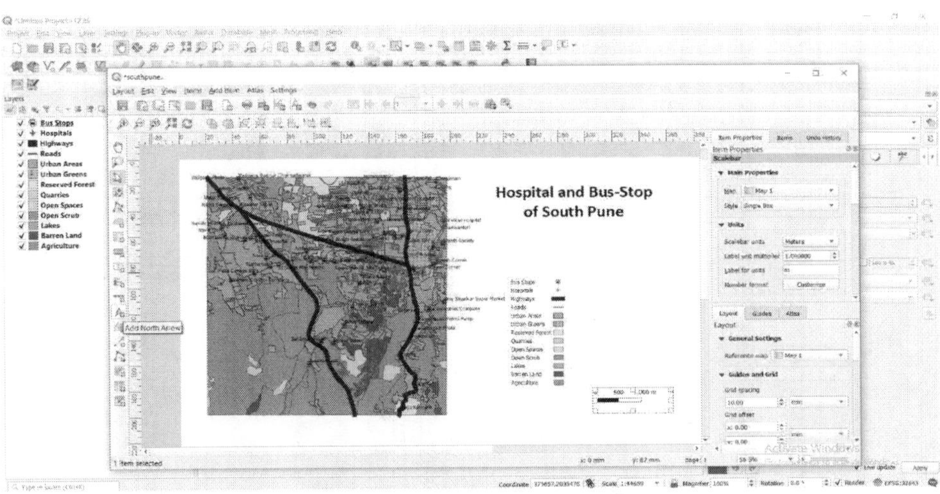

Figure 7.66: Add the north arrow

Step 30: Now click on the centre of the map to select the whole map extent. Now _checkʹ thecheck box of _Frameʹ under the _Item Propertiesʹ. The map will look like below

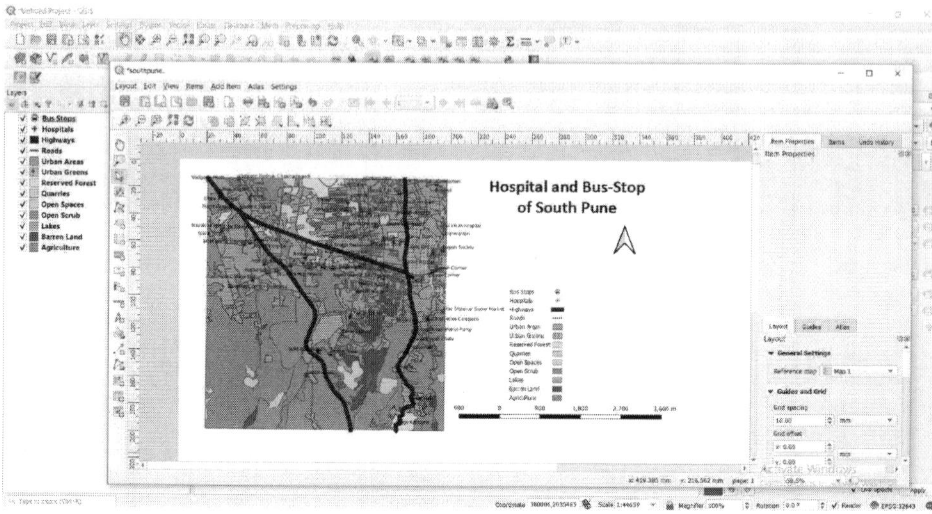

Figure 7.67: Map with all component

Step 31: Now save the map composer by clicking _Saveʹ. Now export the map to a Pdf byclicking the _Export to PDFʹ button. Navigate to a suitable folder, name the file as _SouthPuneʹ and press _Saveʹ

Figure 7.68: Export the map

CHAPTER 8

REVIEW AND EXPLORE THE TUTORIALDATASUPPLIEDIN QGIS AND TOPRESENT IT IN THE FORM OF A THEMATIC MAP

Vinod K kumar Yadav, Palsam Kartik Kumar Goud and Liton Paul

ICAR-Central Institute of Fisheries Education, Mumbai-400061

What you should be able to do after this exercise?

View and explore vector data in QGIS and create a thematic map of vector data.

Principle

1. What is vector data?

Vector data structures represent specific features on the Earth's surface, and assign attributes to those features. These are composed of discrete geometric locations (x, y values) known as vertices that define the shape of the spatial object. It can be a point data, line data, or polygondata.

2. What are thematic maps?

The maps which are used to depict information on a particular topic. This topic can be physical, statistical, measured, or interpreted. Weather maps, population density maps, temperature maps, are examples of thematic maps.

Procedure

Step 1: Open the QGIS Desktop from start menu

Step 2: In Menu bar: Layer → Add Layer → Add Vector Layer

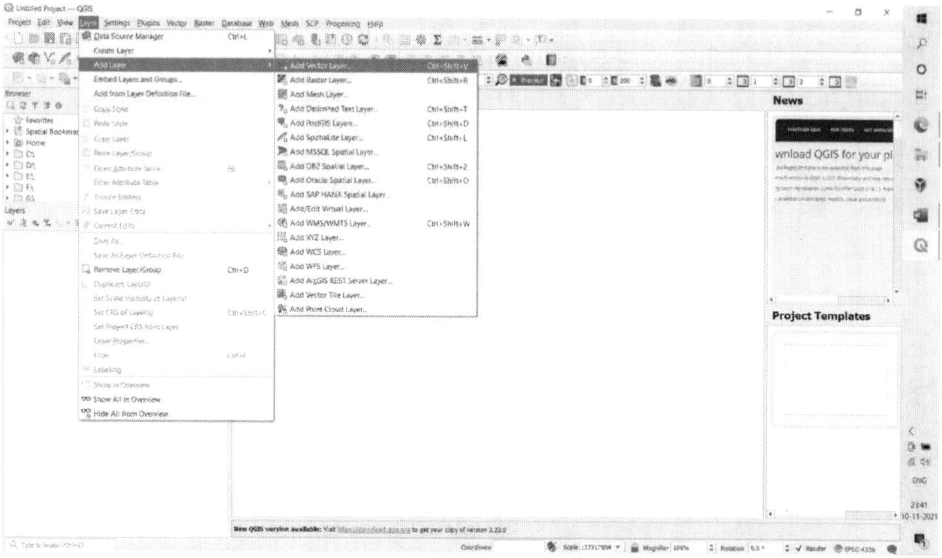

Figure 8.1: Add vector layer option

Step 3: Add the vector Dataset(s) i.e., Pune_beed.shp by Browsing. Pune is on the left and Beed is on the right.

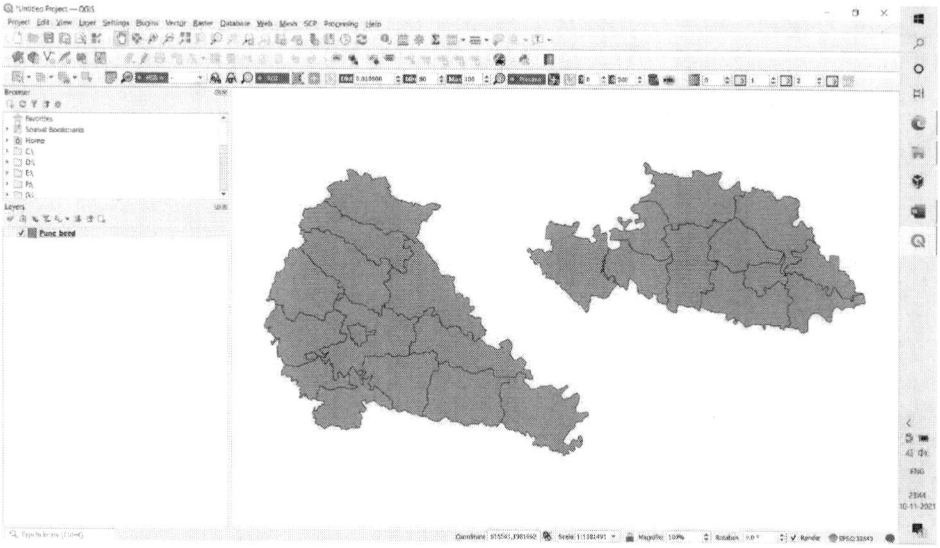

Figure 8.2: Brows the Pune vector data

Step 4: Click on –Identify Features button below the Menu bar or Menu bar: View → Identify Features

Figure 8.3: Identify feature tool from toolbar

(OR)

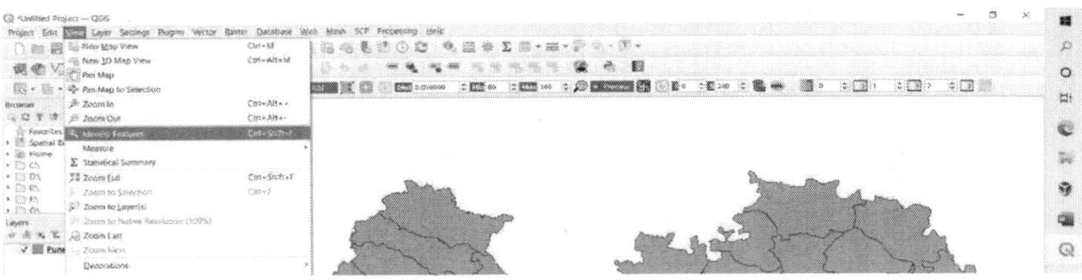

Figure 8.4: identify feature from view tool

Step 5: Click on any Tehsil from Pune or Beed and observe the attributes associated with it. Note that the selected Tehsil gets highlighted.

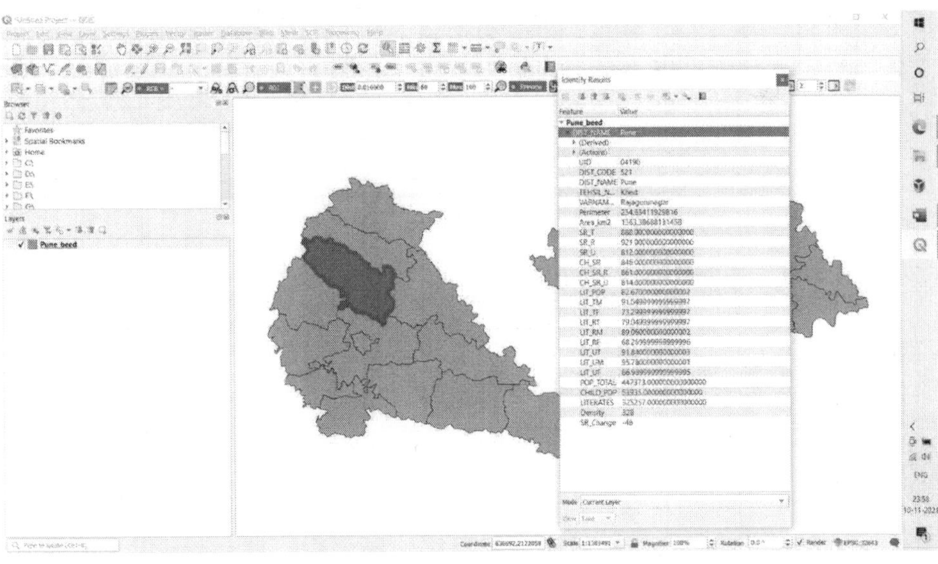

Figure 8.5: Attribute associated with selected feature

Step 6: Close the opened dialogue box, right click on the layer name i.e., Pune_beed in the layer list on the right side of the window. Select –Open Attribute table to view all the attributes in the columns and the features as rows.

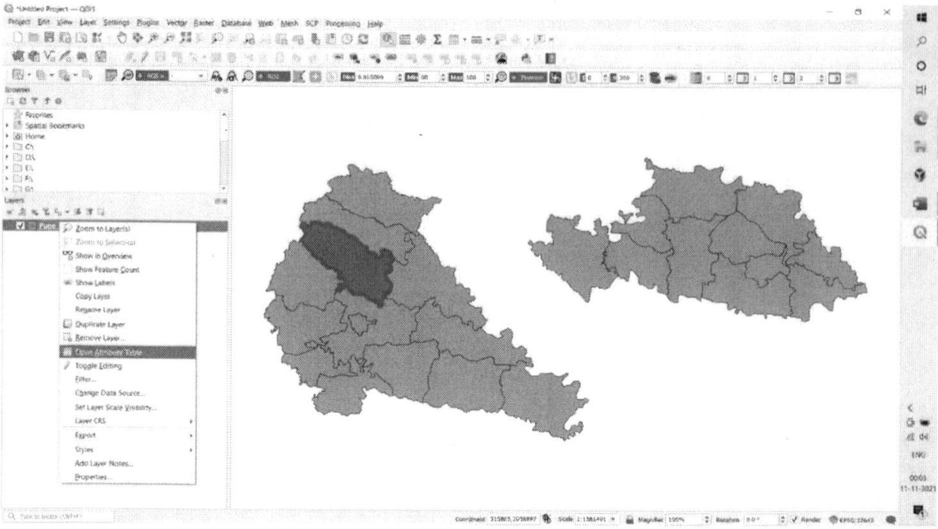

Figure 8.6: Steps to open the attribute table

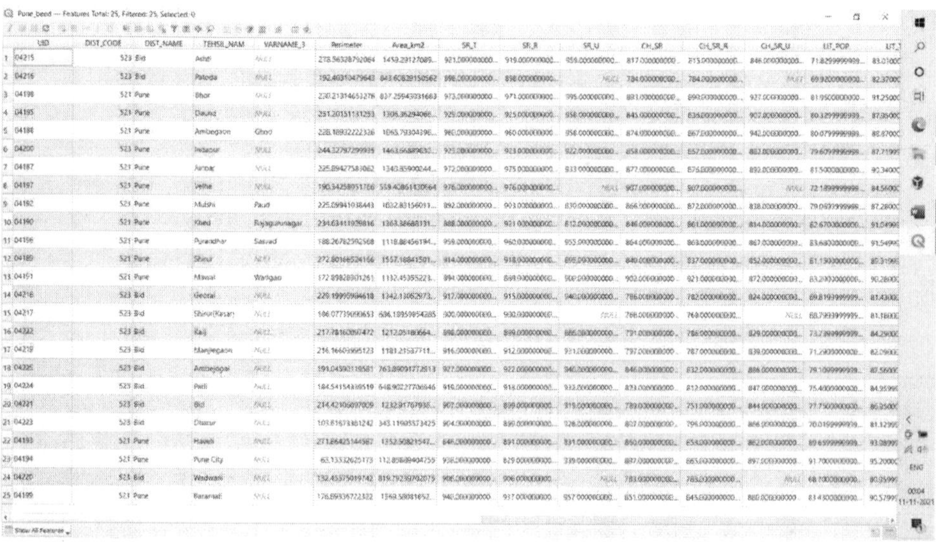

Figure 8.7: Attribute table

Step 7: The attributes can be sorted either ascending or descending order by clicking on the column title. The symbol beside the column title i.e. ▼ and ▲ denote descending and ascending order respectively. The figure below shows data arranged in descending order of –Area_km^2

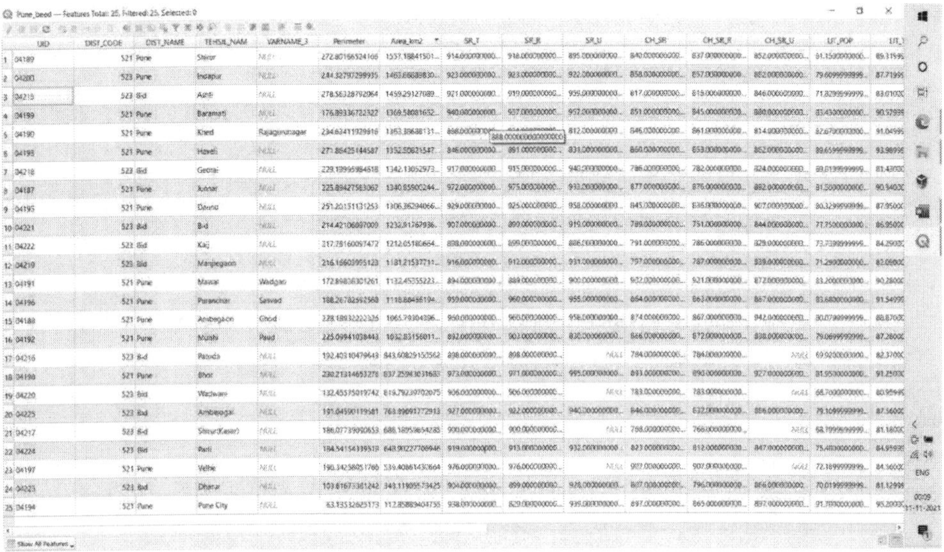

Figure 8.8: Arrangement of the data in descending and ascending order

Step 8: Using column filter to search for a certain feature by attitude. Use –Show All Features dropdown at the bottom, and select –Field Filter or –Column Filter. To select all the Tehsils from Pune district, select –DIST_NAME in the –Field Filter dropdown.

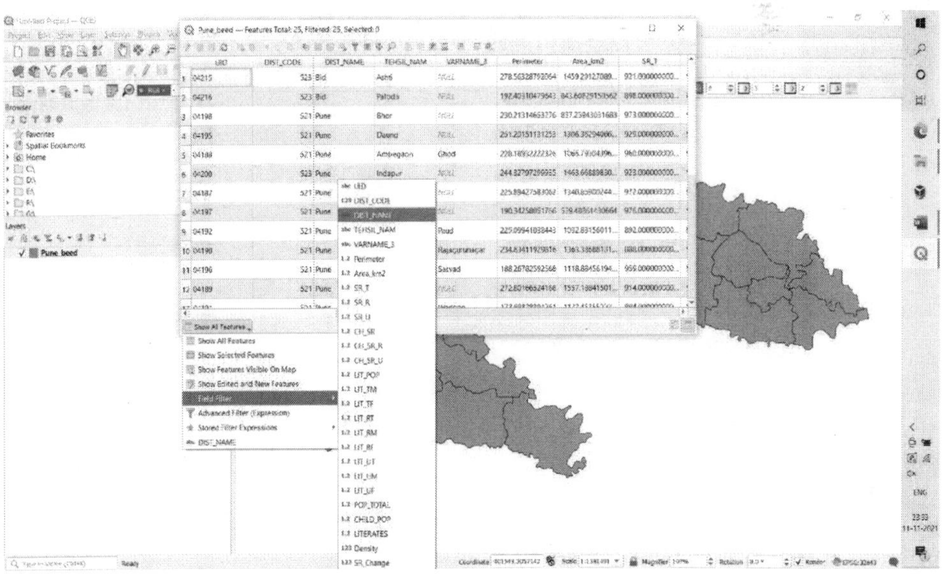

Figure 8.9: Field filter

Step 9: In the search tab, beside the –Show All Features button, type district name as –Pune (Case Sensitive) and click Apply.

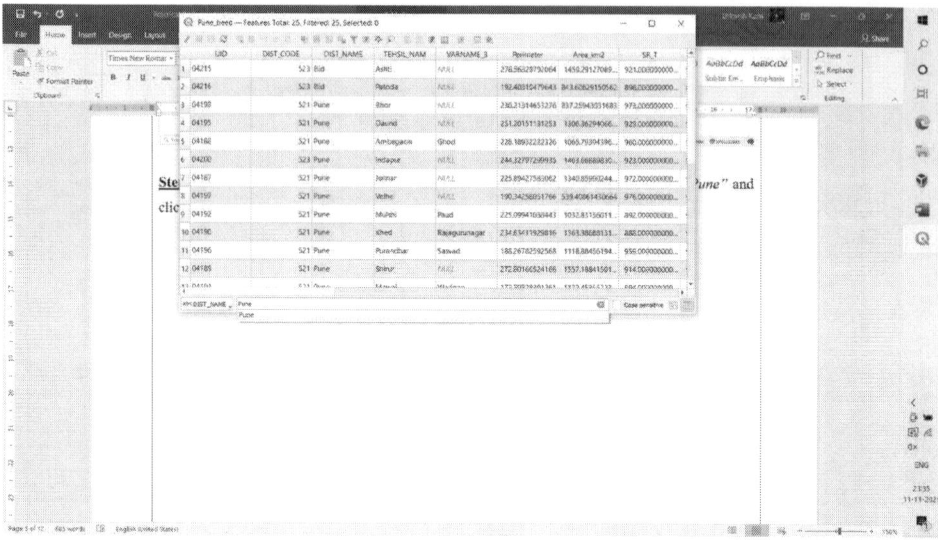

Figure 8.10: Search Pune in search tab

All the Tehsils of only the Pune Tehsil will be shown in the attribute table

Figure 8.11: All the tehsil of Pune district are visible

Step 10: To create a Thematic Map showing Tehsil wise literacy rate, right click on the layer name i.e., Pune_beed in the Layer list on the right side of the window and select –Properties (or simply double click on the layer name).

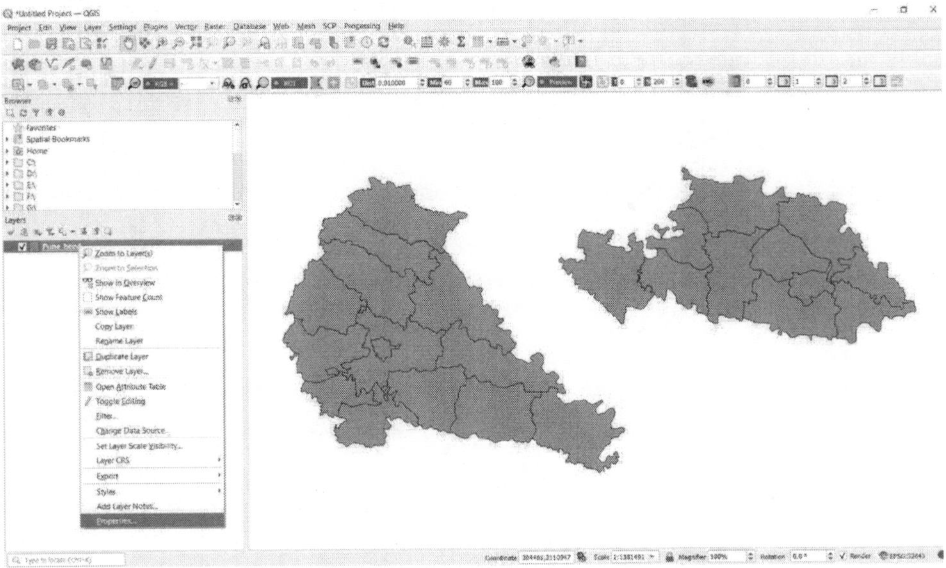

Figure 8.12: Open properties of the vector layer

Step 11: In the Symbology tab, click on -Single Symbol dropdown at top of the window, and change it to -Graduated.

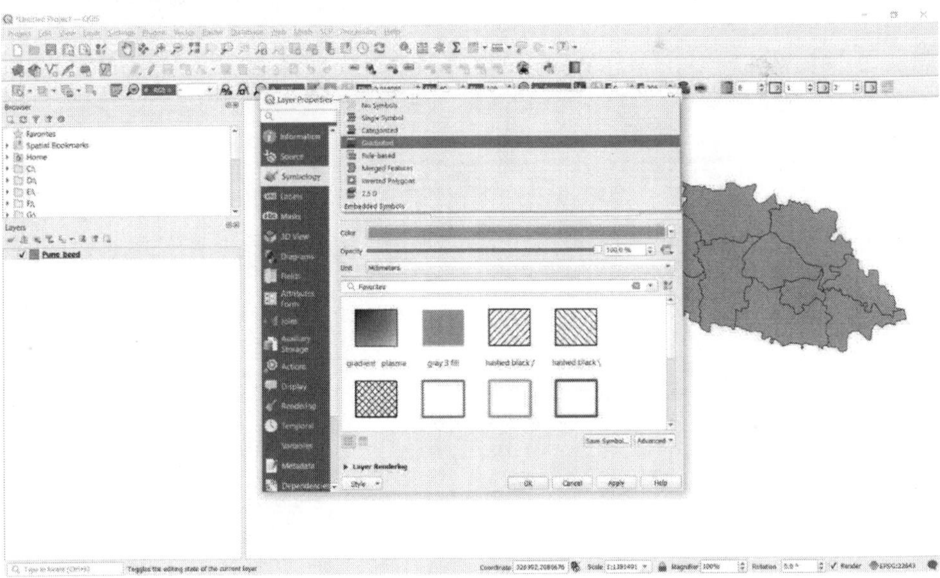

Figure 8.13: Select graduate in symbology dropbox

Step 12: In the –Value dropdown, select –LIT_POP‖ and click OK.

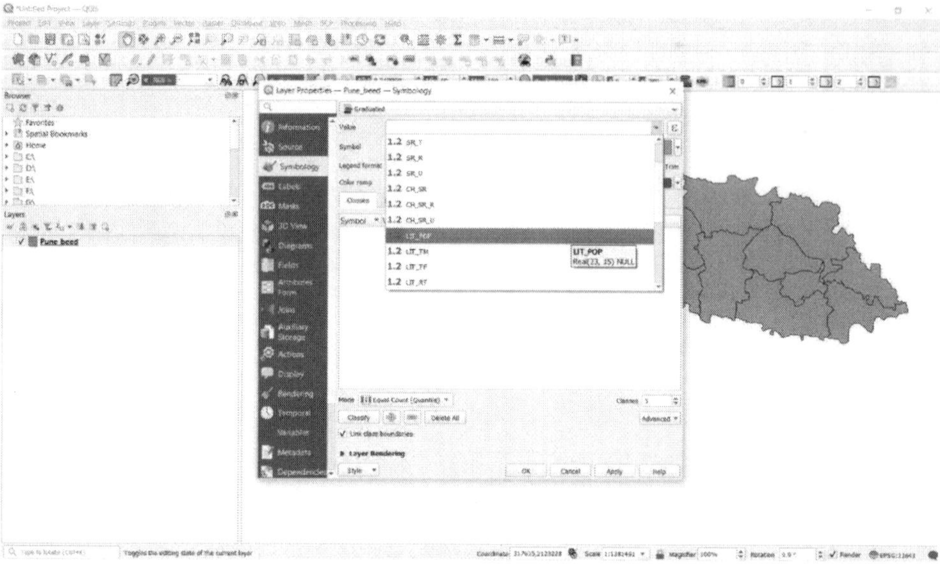

Figure 8.14: Select LIT_POP in the value dropdown

If the classes are not created automatically, click on –Classify button. To change the colors, use –Color Ramp.

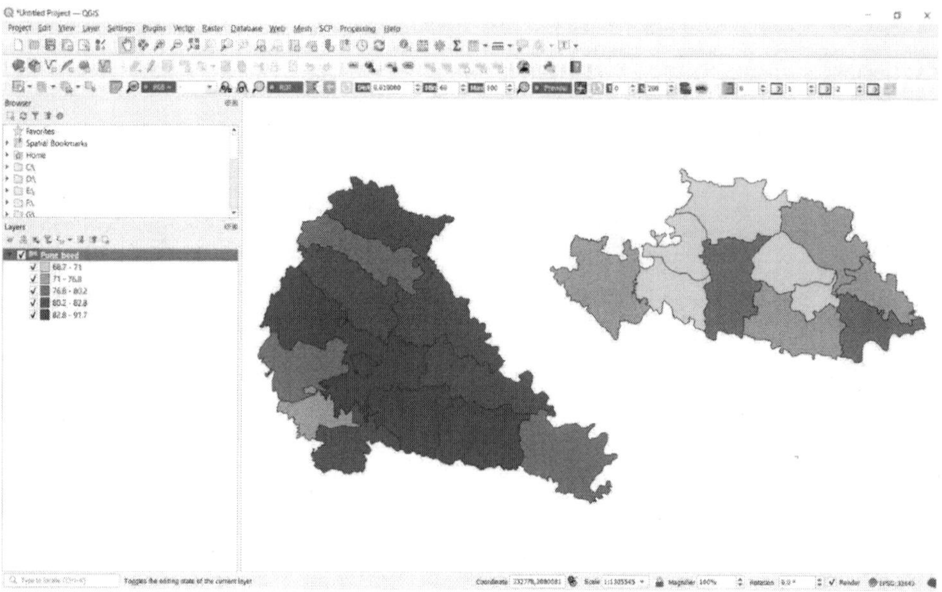

Figure 8.15: Classify and change the color

Step 13: In the same –Properties → –Symbology. _Mode' drop down list can be used to modify class intervals accordingly. For e.g., _Pretty Breaks' mode can be used to create equal intervals.

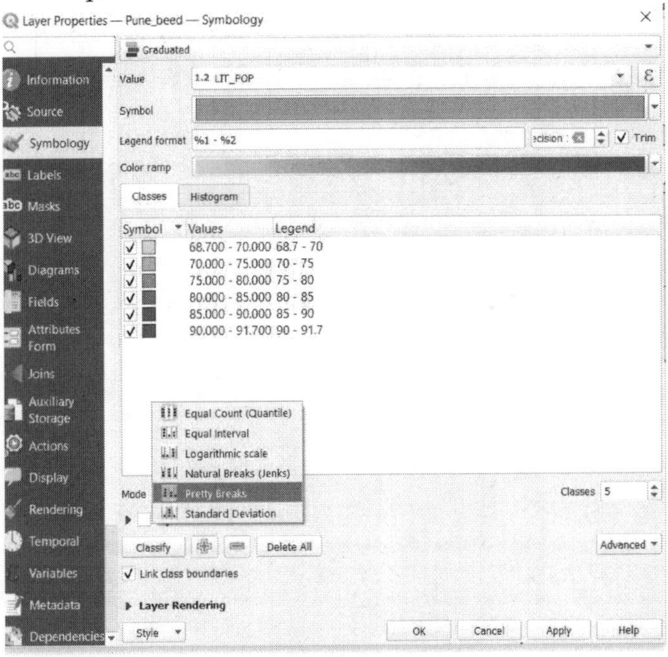

Figure 8.16: Pretty Breaks' mode to create equal intervals

Step 14: To assign own class intervals, double click on the values under the _Value' column and change the lower and upper values accordingly.

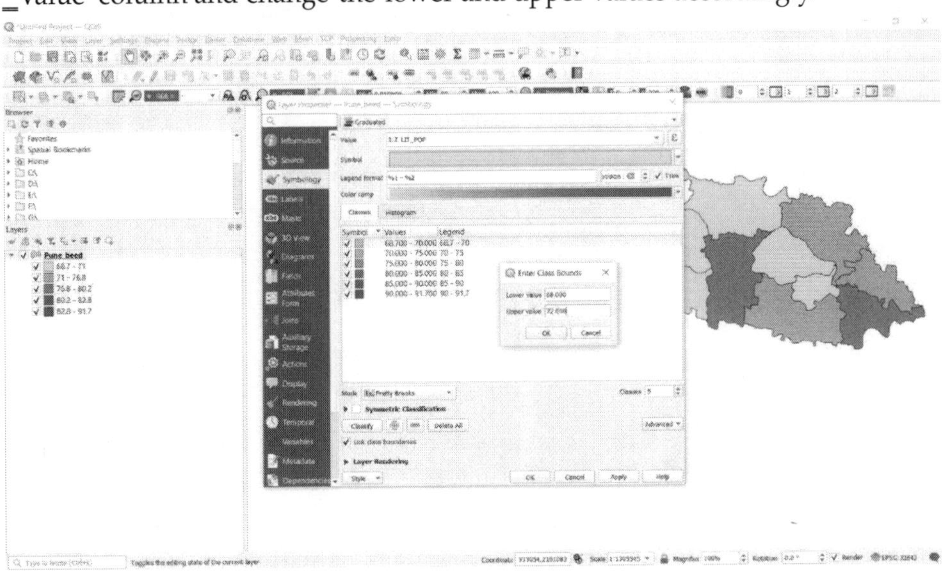

Figure 8.17: Change the value intervals

Step 15: If the class labels have many zeros in the decimal place, double click on the valuesunder Label' beside _Value' and delete the extra zeroes. Then, click on OK.

Step 16: To label the tehsils, open the -Layer Properties window and click on the -Label tab. Click on the –Single Labels option from the dropdown

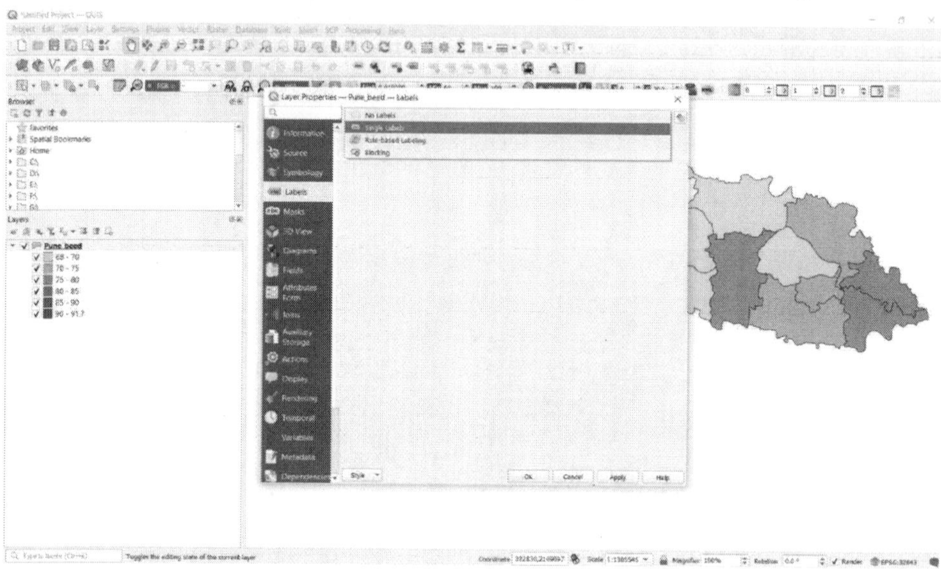

Figure 8.18: In the label chose single label from the dropdown

Step 17: In the –Single Labels option, chose –TEHSIL_NAM from the dropdown of –Value option. Choose the required format of Label and click on OK.

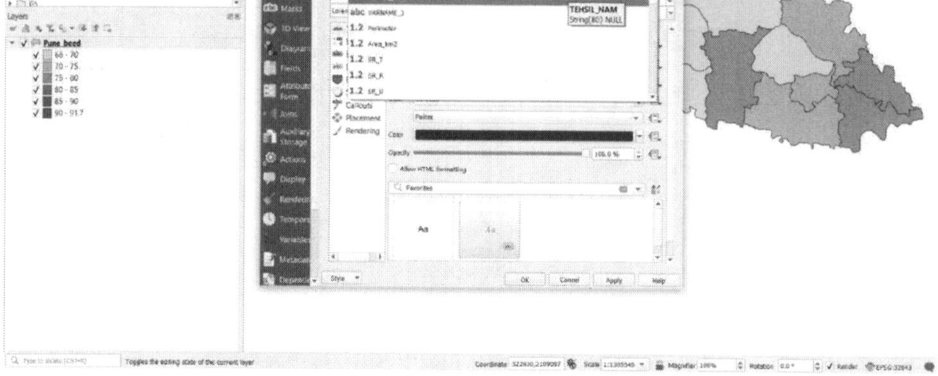

Figure 8.19: Select TEHSIL_NAM

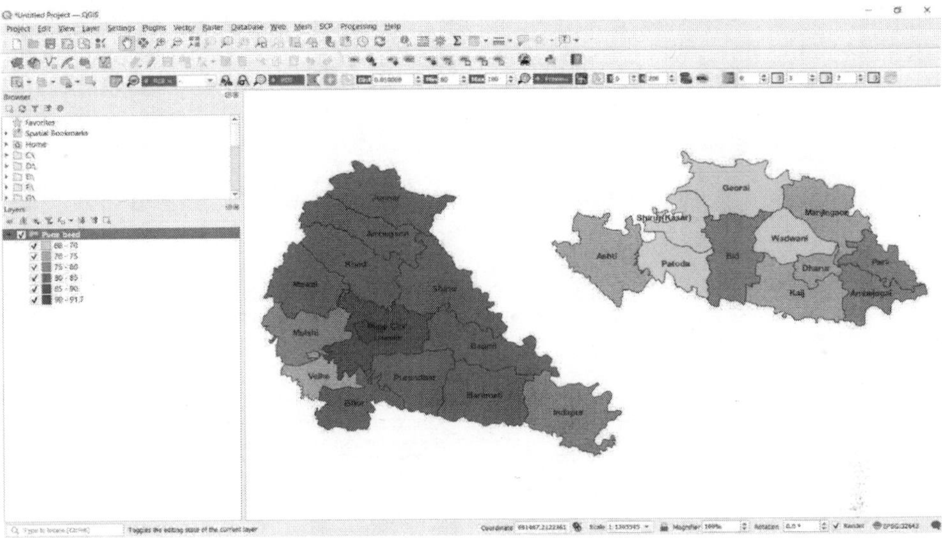

Figure 8.20: Label of the polygones

Displaying output as MAP

Step 1: In Menu bar, Project New Print Composer/ New Print Layout

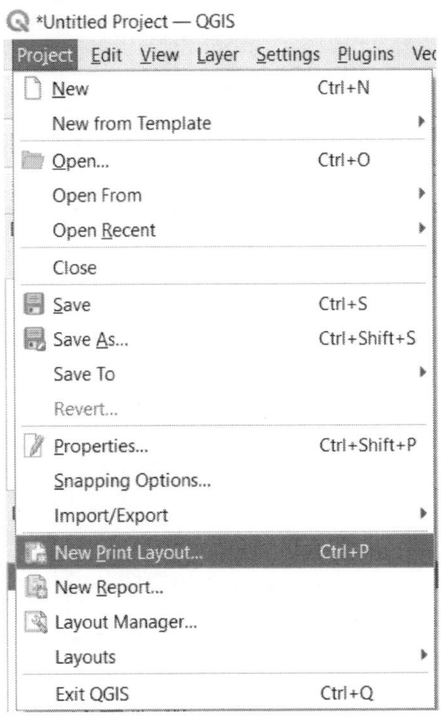

Figure 8.21: New print layout from the project to prepare the map

Step 2: The new dialog box will open up asking for the title of composed. Give any title asrequired, in this case, Pune_Beed. Then, click on OK.

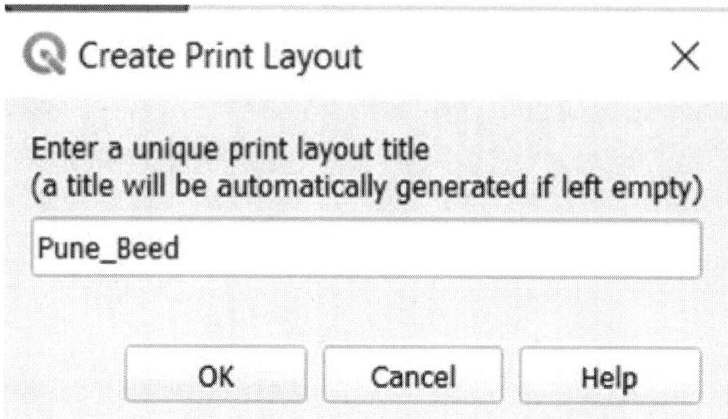

Figure 8.22: Give title of composed

A Blank Map Canvas is opened

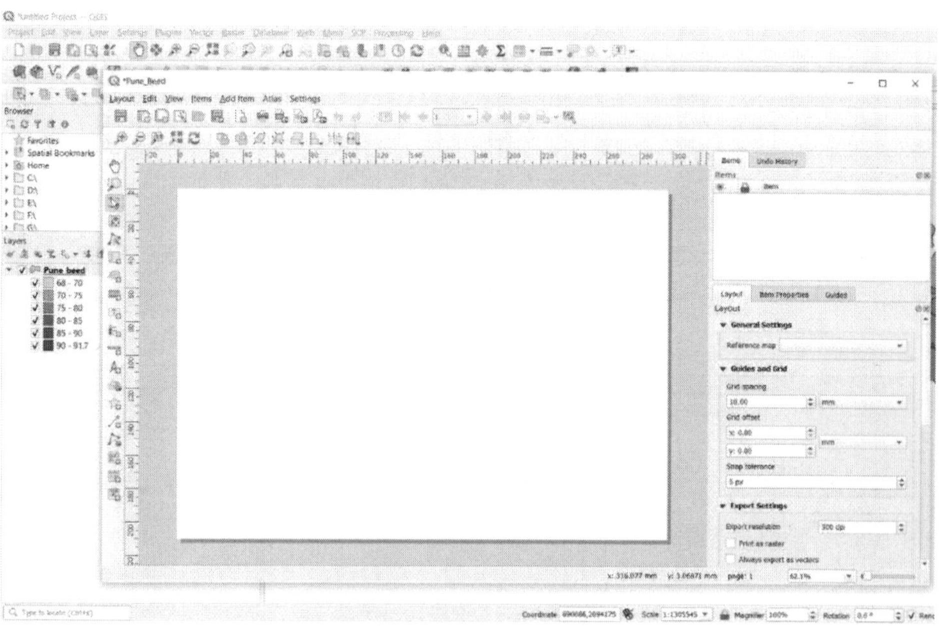

Figure 8.23: Blank map

Step 3: In the Menu Bar of the Map Canvas, add the map to the canvas by selecting, AddItem → Add Map

Figure 8.24: Add map

Step 4: Click on one corner of the canvas and drag out the desired extent of the map.

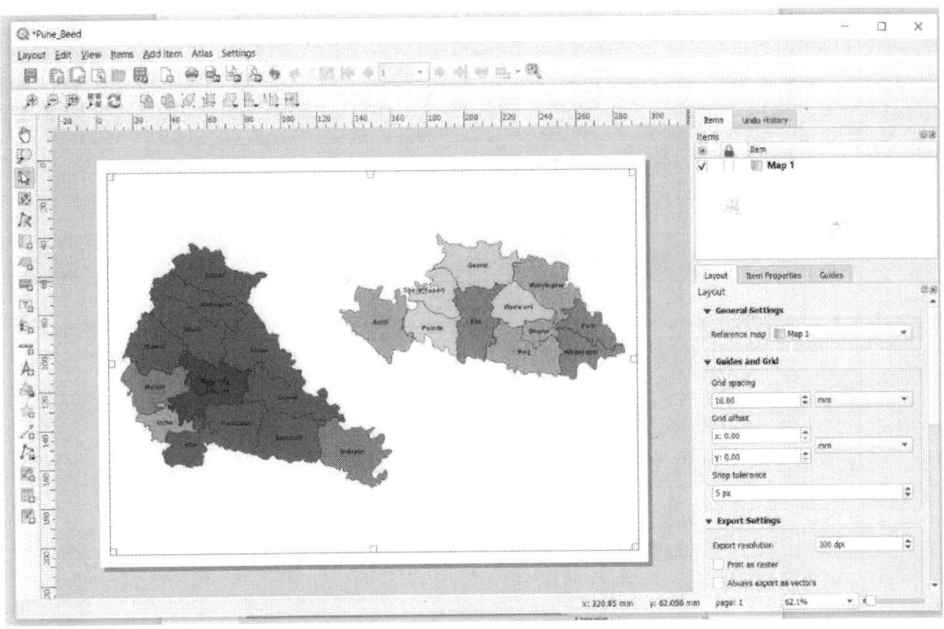

Figure 8.25: Adjust the corners of the map by dragging

Step 5: To change the scale of the layer, click on the map box and then on the –Item Properties tab. In the scale option, type the scale accordingly. In this case, let it be 1500000.

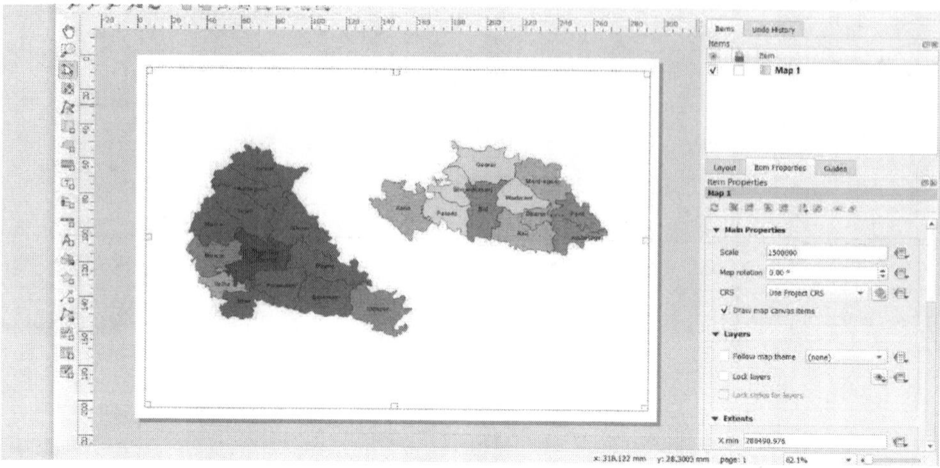

Figure 8.26: Change the scale

Step 6: In Menu bar, under –Add Item, add Legend, North Arrow, Scale Bar, and Title to prepare a map.

To place these items, click on its button and then click where to place it on the map. To adjust its properties, select the item and change the setting via Item Properties' tab.

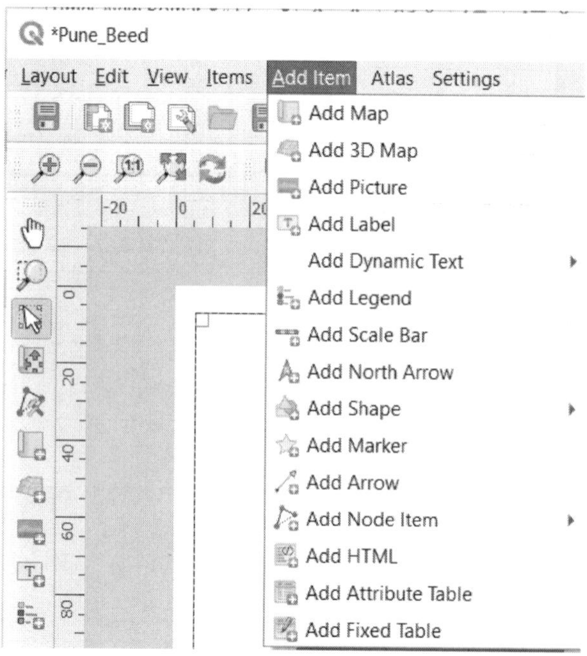

Figure 8.27: List of the items can be added to the map

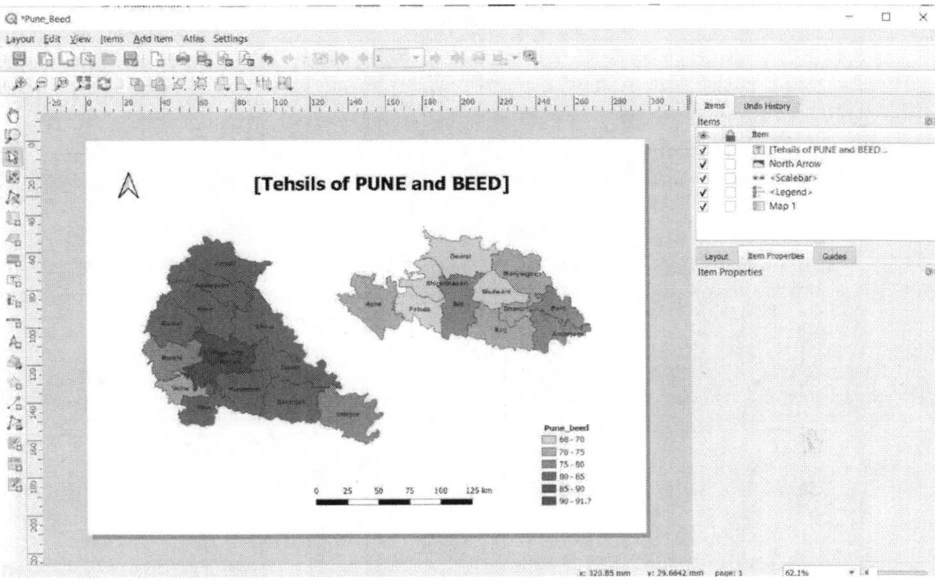

Figure 8.28: Final map with title, legend and scale bar

Step 7: In the Menu bar, select –Layout and then –Export to PDF to print out the map as pdf.

Observations

1. Labels and Colour themes can be assigned according to requirement.
2. Map can be prepared according to the requirement, and with any required formatting.

CHAPTER 9

CONSTRUCTING SPATIAL QUERIES IN QGIS

Vinod K kumar Yadav, Palsam Kartik Kumar Goud and Liton Paul

ICAR-Central Institute of Fisheries Education, Mumbai-400061

What you should be able to do after this exercise?

Perform spatial queries on various features based on location and relationships.

Data: Exercise_7

Principle

1. What is a spatial relation?

A connection between spatial features concerning their spatial arrangements and spatial locations. Spatial relations can be of three types: proximity relations, topological relations and direction relations.

2. What are spatial queries?

A spatial query selects features based on their spatial relationships to other features and areused to answer spatial questions

Procedure

Query 1: Find out the road segments in Maharashtra where road cross over the river.

Data used: Shape files- road_maha.shp' and River_Maha.shp'

Step 1: Open the QGIS Desktop from start menu

Step 2: In Menu bar: Layer → Add Layer → Add Vector Layer

Step 3: Add the vector Dataset(s) i.e., road_maha.shp and River_Maha.shp by browsing.

Step 4: To perform spatial query by location, Go to Vector → Research Tools → Select byLocation

Or

Processing → Toolbox

Type ‗Select by location' and chose the tool

Step 5: In the window opened, Select Features from: Select ‗ road_maha' from the drop-downlist

Where the features (geometric predicate): Check the box beside ‗cross' By Comparing to the features from: Select River_Maha'

Modify current selection by: creating new selectionClick on Run

Step 6: The process takes some time to run.

Step 7: Close the window. You will notice that all the corresponding features are highlightedin yellow colour.

Step 8: To save the selected features as separate layer. Right click on the road_maha' layerin the Layer Bar (in the left) and Export ‗Save Selected Features

Step 9: Save the layer with Format –ESRI Shapefile

Step 10: To move to next query, deselect the selected features.

On selection toolbar, use Deselect Features from all Layers' option

Query 2: Find out the river channels that are not flowing through Ahmadnagar District.

Data used: Shape files- ‗Maha_Tehsil.shp' and ‗River_Maha.shp'

Step 1: Open the QGIS Desktop from start menu

Step 2: In Menu bar: Layer → Add Layer → Add Vector Layer

Step 3: Add the vector Dataset(s) i.e., Maha_Tehsil.shp and River_Maha.shp by browsing (keep the River_Maha layer on top of Maha_Tehsil layer)

Step 4: Since we want rivers flowing outside Ahmadnagar, firstly we will select Ahmadnagardistrict from the Maharashtra state. This can be done by using attribute query and we will perform spatial query on the selected data from attribute query. Right click on ‗Maha_Tehsil' and select ‗Filter'.

Step 5: The query builder window gets opened. In this window, select all the tehsils of Ahmadnagar district by using attribute query, i.e., double click on "NAME_2" from fields window and clickon clickon‗=' in the ‗Operators' section

Step 6: Then to see all the values in Name_2 field click on All' in Values' section in window and select Ahmednagar from values

Step 7: In short, we need to create an expression "NAME_2" = 'Ahmednagar' and Click OK'.

Step 8: This will select all the tehsils from Ahmednagar district.Step 9: To perform spatial query by location.

Go to Vector → Research Tools → Select by Location

Or

Processing → Toolbox

Type ‗Select by location' and chose the tool

Step 10: In the window opened, select Features from: Select ‗River_Maha' from the dropdown list Where the features (geometric predicate): Check the box beside disjoint'

By Comparing to the features from: Select ‗Maha_Tehsil'And use the result to: Create new selection

Click Apply

Click on Run

Step 11: The process takes some time to run.

Step 12: The river channels which are not flowing through Ahmadnagar would behighlighted in yellow color in map canvas.

Step 13: To save the selected features as separate layer. Right click on the River_Maha'layer in the Layer Bar (in the left) and Export → Save Selected Features

Step 14: Save the layer with Format –ESRI Shapefile ‖

Step 15: To move to next query, deselect the selected features. On selection toolbar, use ‗Deselect Features from all Layers' option

Query 3: Find out the public places that are within Greater Mumbai.

Data used: Shape files- ‗Maha_Tehsil.shp' and ‗public_place_maha.shp' Step 1: Open the QGIS Desktop from start menu

Step 2: In Menu bar: Layer → Add Layer → Add Vector Layer

Step 3: Add the vector Dataset(s) i.e., Maha_Tehsil.shp and public_place_maha.shp byBrowsing (keep the public_place_maha layer on top of Maha_Tehsil layer)

Step 4: Since we want public places inside Greater Mumbai, firstly we will select GreaterBombay district from the Maharashtra state.

This can be done by using attribute query and we will perform spatial query on the selecteddata from attribute query.

Right click on Maha_Tehsil' and select Filter'.

Step 5: The query builder window gets opened. In this window, select all the tehsils of Greater Bombay district by using attribute query, i.e., double click on "NAME_2" from fields window and click on click on_ =' in the_ Operators' section. Then to see all the valuesin Name_2 field click on _All' in_ Values' section in window and select Greater Bombay from values

Step 6: This will select all the tehsils from Greater Bombay

Step 7: To perform spatial query by location.

Go to Vector → Research Tools → Select by Location

Or

Processing → Toolbox

Type _Select by location' and chose the tool

Step 8: In the window opened, select Features from: Select _public_place_ maha' from the drop down list. Where the features (geometric predicate): Check the box beside _are within'

By Comparing to the features from: Select _Maha_Tehsil'. And use the result to: Create newselection

Click Apply

Click on Run

Step 9: The process takes some time to run. The public places which are present in theGreater Bombay region gets highlighted in yellow

Step 10: To save the selected features as separate layer. Right click on the _public_place_maha' layer in the Layer Bar (in the left) and Export → Save SelectedFeatures.

Step 11: Save the layer with Format –ESRI Shapefile ll

Step 12: To move to next query, deselect the selected features. On selection toolbar, use

Deselect Features from all Layers' option

Query 4: Find out the schools in Maharashtra that could be affected by floods?

Let assume the flood zone is at about 3 km from the river bank.

Data used: Shape files- _School_col_Maha.shp' and _River_Maha.shp'Step 1: Open the QGIS Desktop from start menu

Step 2: In Menu bar: Layer → Add Layer → Add Vector Layer

Step 3: Add the vector Dataset(s) i.e., School_col_Maha.shp and River_Maha.shp bybrowsing

Step 4: To find out Rivers that could be affecting the schools in case floods, firstly we haveto create a flood zone around the rivers with 3 Km by using Buffer function.

To create a buffer around the river, Menu bar → Vector → Geoprocessing tools → Buffer... Step 5: The _Buffer...' window will open. Select Input vector layer as River_Maha'.Specify _Buffer Distance' as 3km (Since our working unit is meter in UTM projection). Restof the parameters same. Save the output shapefile (in desired location in local drive with aproper name). This can be done with the help of browse button. Once the inputs are givenproperly click on _OK'. If you asked to add the output layer to map canvas, click on yes.

Close the Buffer window and the output layer looks as follows:

Move the _School_col_Maha' layer on the top of this newly created buffer layer.Step 6: To perform spatial query by location.

Go to Vector → Research Tools → Select by Location

Or

Processing → Toolbox

Type _Select by location' and chose the tool

Step 7: In the window opened, select Features from: Select _School_col_Maha' from thedrop down list. Where the features (geometric predicate): Check the box beside _are within' By Comparing to the features from: Select Buffer'

And use the result to: Create new selection

Click Apply

Click on Run

Step 8: The process takes some time to run. The schools which are present in the 3 KMregion to the rivers gets highlighted in yellow

Step 9: To save the selected features as separate layer. Right click on the _School_col_maha'layer in the Layer Bar (in the left) and Export > Save Selected Features

Step 10: Save the layer with Format –ESRI Shapefile ||

Step 11: To move to next query, deselect the selected features. On selection toolbar, use

_Deselect Features from all Layers' option

Query 5: Find out the tehsils in Pune which contains water bodies such as lakes andDams?

Data used: Shape files- _waterbodies_maha.shp' and _Maha_Tehsil.shp'Step 1: Open the QGIS Desktop from start menu

Step 2: In Menu bar: Layer → Add Layer → Add Vector Layer

Step 3: Add the vector Dataset(s) i.e., waterbodies_maha.shp and Maha_Tehsil. shp by Browsing (Keep the waterbodies_Maha layer over Maha_Tehsil)

Step 4: Since we want water bodies inside Pune, firstly we will select Pune district from the Maharashtra state.

This can be done by using attribute query and we will perform spatial query on the selected data from attribute query. Right click on Maha_Tehsil' and select _Filter'.

Step 5: The query builder window gets opened. In this window, select all the tehsils of Pune district by using attribute query, i.e., double click on "NAME_2" from fields window and click on click on_ =' in the_ Operators' section.

Then to see all the values in Name_2 field click on _All' in_ Values' section in window and select Pune from values

In short, we need to create an expression "NAME_2" = 'Pune' and Click _OK'. Step 6: This will select all the tehsils from Pune.

Step 7: To perform spatial query by location.

Go to Vector → Research Tools → Select by Location

Or

Processing → Toolbox

Type _Select by location' and chose the tool

Step 8: In the window opened, select Features from: Select _Maha_Tehsil' from the dropdown list

Where the features (geometric predicate): Check the box beside _contain'. By Comparing tothe features from: Select _waterbodies_Maha'

And use the result to: Create new selection

Click Apply

Click on Run

Step 9: The process takes some time to run. The tehsils of Pune which contain water bodiesget highlighted in yellow.

Step 10: To save the selected features as separate layer. Right click on the _Maha_tehsil'layer in the Layer Bar (in the left) and Export → Save Selected Features

Step 11: Save the layer with Format –ESRI Shapefile ‖

Step 12: deselect the selected features. On selection toolbar, use _Deselect Features from allLayers' option.

Constructing spatial queries in QGIS

What you should be able to do after this exercise?

Able to query spatial data in QGIS

Principle

1. What is vector data?

Vector data structures represent specific features on the Earth's surface, and assign attributes to those features. These are composed of discrete geometric locations (x, y values) known as vertices that define the shape of the spatial object. It can be a point data, line data, or polygondata.

2. What is spatial query?

Spatial query refers to the process of retrieving a data subset from a map layer by working directly with the map features. In a spatial database, data are stored in attribute tables and feature/spatial tables.

3. What is attribute?

Attribute. [data models] Non spatial information about a geographic feature in a GIS, usuallystored in a table and linked to the feature by a unique identifier. For example, attributes of a river might include its name, length, and sediment load at a gauging station.

Procedure

Step A: Open the QGIS Desktop from start menu

Step B: In Menu bar: Layer → Add Layer → Add Vector Layer

Step C: Add the vector data provided (Shape files).

Query 1: From the given data-sets, which tehsil in two districts have least population?

Step 1: Right click on the layer Pune_Beed_POP.shp → Open Attribute Table.

Step 2: Search for the –POP_TOTAL || column in the attribute table.

Step 3: Triangle pointing upward indicates that the data in the column is in 'Ascending order' and vice-versa. Arrange the data in ascending order for 'POP_TOTAL'. So the first entry will have minimum population, which we are looking for.

Step 4: Click on the top number at the left most corner of the table.

Step 5: To create a new shape file of the selected area, right click on layer name i.e., 'Pune_beed_POP.shp' under Export and click on _Save Selection As'

Step 6: You will be directed to 'Save vector layer as' window. Select desired name and location using 'Browse' click Ok to save.

Query 2: Find out the areas having child sex ratio ("CH_SR") above 850.

Step 1: Open the Attribute table of Pune_beed_POP.shp'. Click on _Show All Feature' tab at the left bottom of the table and then click on _Advance Filter (Expression)'.

Step 2: Expression based Filter' window will be open up. This window contains different functions and list of fields.

Step 3: Type the Desired expression for your query to get the output here we will use CH_SR (Chile Sex Ration)>65 as expression and click on Ok to get the result after getting the result we need to select the features to be highlighted on the existing map.

Step 4: This condition will select all the areas having child sex ratio greater than 850. Select these areas to highlight it on the map. As discussed earlier we can export these selected areasas separate vector layer to create a map.

Query 3: Which are areas having sex ratio below 950 and total female literacy below 65%.

Step 1: Open the Attribute table of _Pune_beed_POP.shp'. Click on _Show All Feature' tab at the left bottom of the table and then click on _Advance Filter (Expression)'.

Step 2: Expression based Filter' window will be open up. This window contains different functions and list of fields. Type the desired expression in the expression field here the expression is –SR_T ‖ < 950 AND –LIT_TF ‖ < 65.

Step 3: This will select all the areas having sex ratio less than 950and literacy below 65%.

Query 4: Find out the areas having Latrine facility more than 40% and literacy rategreater than 75%.

Step 1: Join –Pune_beed_LAT ‖ and –Pune_beed_POP ‖ with the help of common field as UID.

Step 2: Now both the attributes required for the query solution are there in a single table. So,open the attribute table of 'Pune_beed_LAT'; Go in Advance Filter (Expression)' and enter

("latperhous" >40) AND ("Pune_beed_POP_LIT_POP" >75)' equation in expressionwindow.

Step 3: Input the expression to get the Latrine facility>40 and literacy rate greater than 75%.Step 4: Apply the expression to get the query answered

Step 5: Select the output query to get the area highlighted on the map.

Step 6: The selected feature can be exported as new shape file from layer → Export → Saveselected feature.

CHAPTER 10

LULC SUPER CLASSIFICATION THROUGH MACHINE LEARNING TECHNIQUE OR GEE PLATFORM

Vinod Kumar Yadav[1], Chandani Dave[1] and Mohammed Meharoof[1,2]

1 ICAR-Central Institute of fisheries Education, Mumbai-400061
2 Assistant Professor (o/c), College of Fisheries, Payyanur, Kannur-670307

Sign-up for Google Earth Engine

Introduction

Google Earth Engine is a powerful geospatial platform developed by Google that allows users to analyze and visualize vast amounts of Earth observation data for a wide range of applications, from environmental monitoring to urban planning.

New registration

Search Google Earth engine Code editor in the browser. Get Started Select the Mail ID with which you want to register

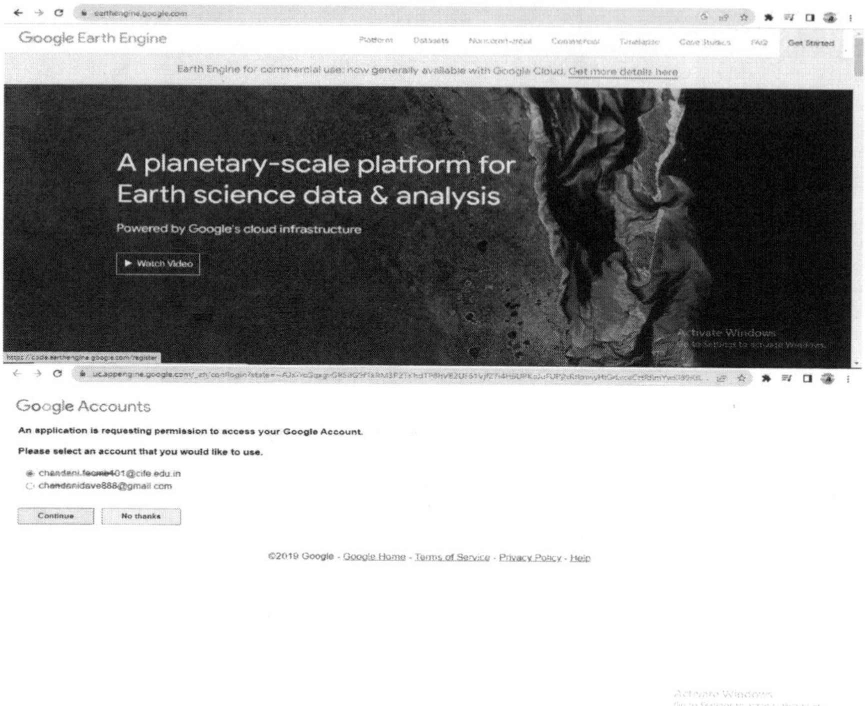

Figure 10.1: Registration on the Google Earth Engine

Click on the link given for the noncommercial users for sign in

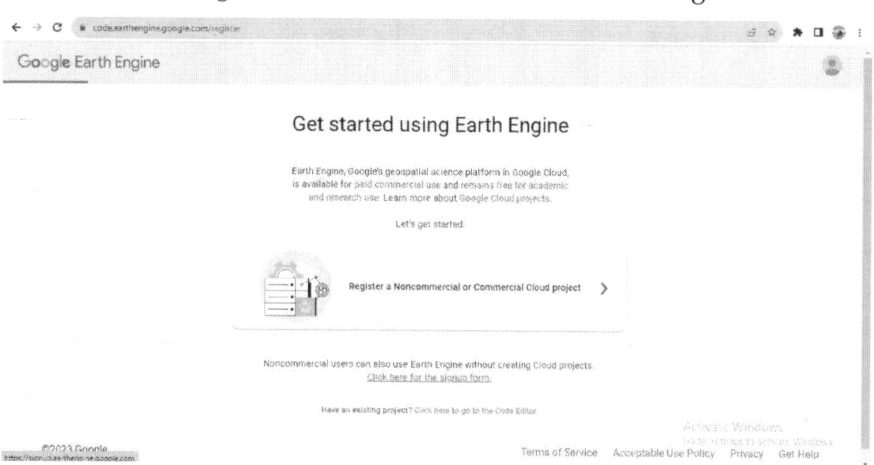

Figure 10.2: Click on the link – register a noncommercial or commercial cloud project

On clicking the link one sign-up page will pop up.Fill in all the sign-up details and submit the form.

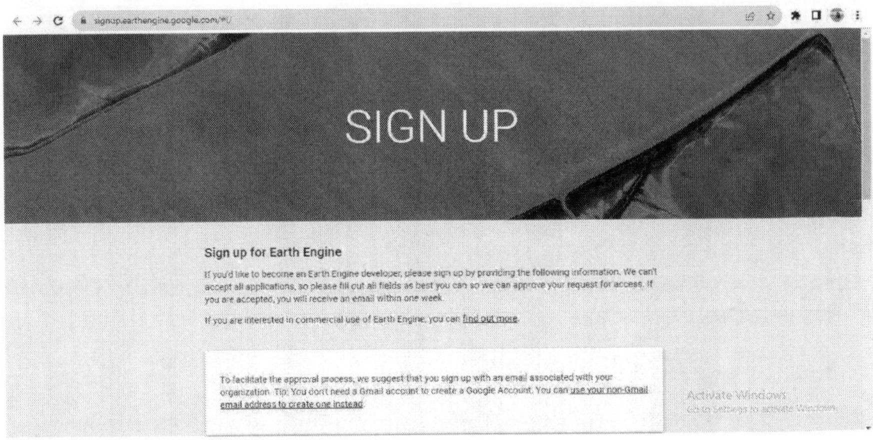

Figure 10.3: Sign up page

Figure 10.4: Sign up details

After submission, one note will come like –Keep an eye on your inbox for xxxxxxxxxxx@gmail.com. We'll let you know when you're good to go.‖

Check your inbox for the mail from the GEE team. Mail may take up to one week to consider your request.

How to Add Shape File to the GEE Code Editor

Prepare Your Shapefile:

Ensure your shapefile is correctly formatted and contains the geographic data you want touse in GEE.

Create Folder in Assets: Access GEE code editor.

In assets create one folder where you can upload your shape file

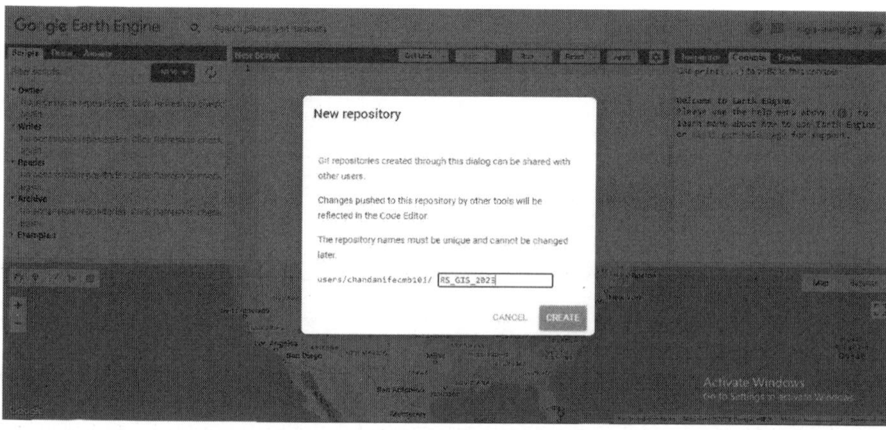

Figure 10.5: Folder creation in asset section

Click New tab in Assets and select Shapefiles

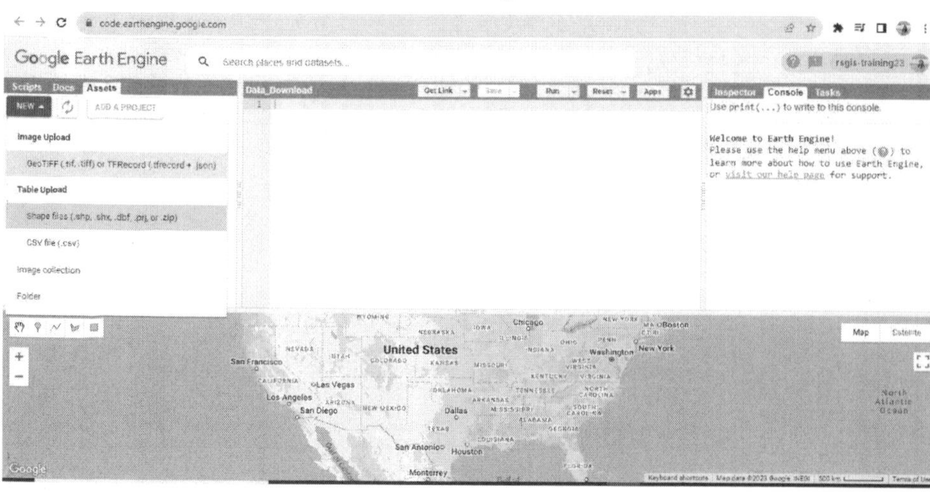

Figure 10.6: Navigating to add the shape files

Click Select in source files upload shp, zip, dbf, prj, shx, cpg, fix, qix, sbn or shp.xml.files Given name in Asset ID Click Upload

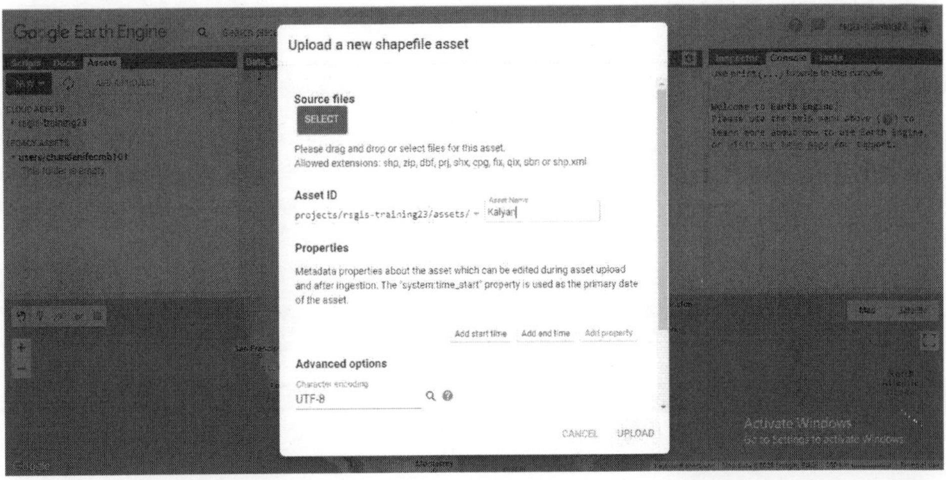

Figure 10.7: Click on select and brows the shape file to be upload

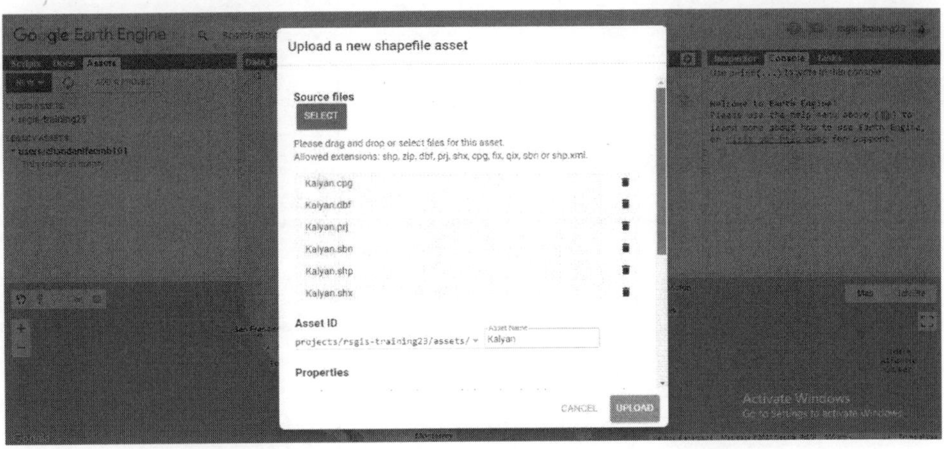

Figure 10.8: Select the files with the mention extensions

Successfully upload shape file will be visible in assets.

Click on the –arrow on the right side of the name to add the shape file into a script.

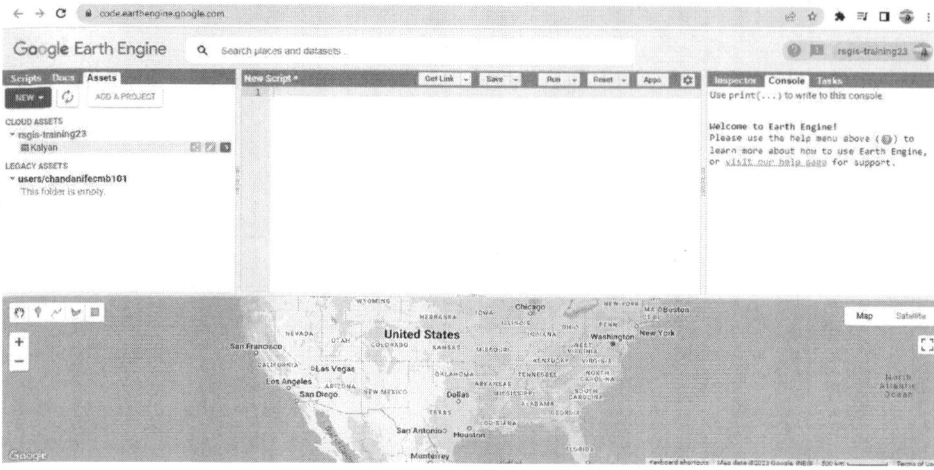

Figure 10.9: Click on the arrow on the right side of the kalyan to add the shape file on the canvas

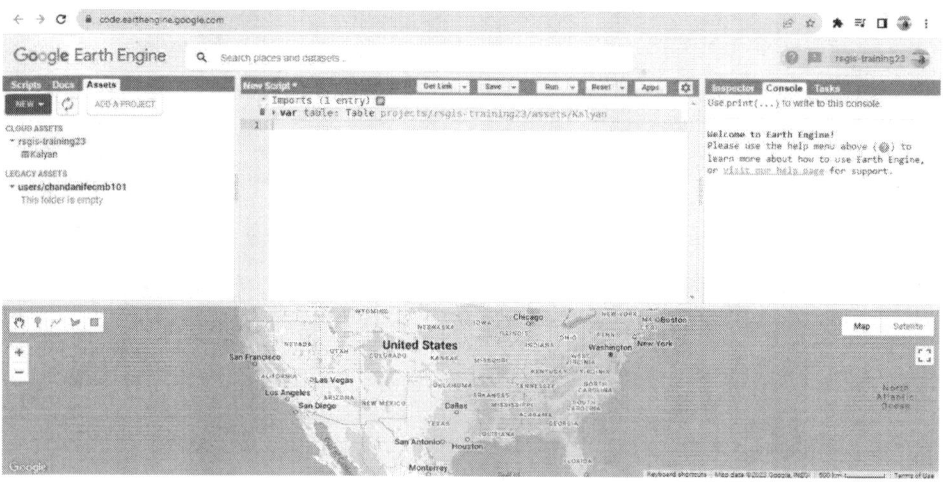

Figure 10.10: Shape file added to the code editor

CHAPTER 11

DATA REPOSITORY OF GEE

Vinod Kumar Yadav[1], Chandani Dave[1] and Mohammed Meharoof[2]

[1] ICAR-Central Institute of fisheries Education, Mumbai-400061
[2] Assistant Professor (o/c), College of Fisheries, Payyanur, Kannur-670307

Introduction

The Google Earth Engine (GEE) data repository is a vast and continually expanding collection of geospatial datasets from various sources, including satellites and ground-based measurements. This repository is a valuable resource for researchers, scientists, and analysts, providing easy access to petabytes of Earth observation data for applications in climate research, land cover analysis, disaster monitoring, and more. With GEE's data repository, users can harness the power of this rich data archive to conduct geospatial analyses and gain insights into our changing planet.

GEE offers a range of satellite imagery from various missions, including Landsat, Sentinel, MODIS, and more. These datasets capture Earth's surface at different spatial and temporal resolutions.

Download the satellite data

Click on the name of the data set you want to download, i.e., Sentinel data

Scroll down the window and select the type of processed data you want to download

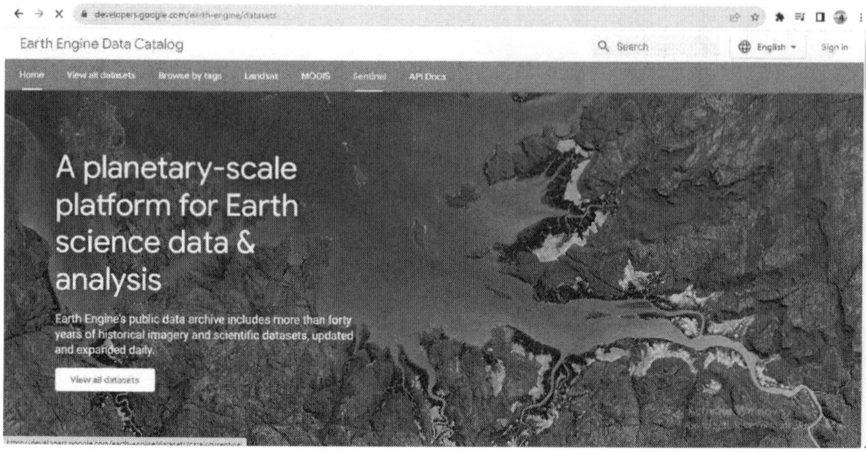

Figure 11.1: Select dataset from the options given, i.e., Sentinel

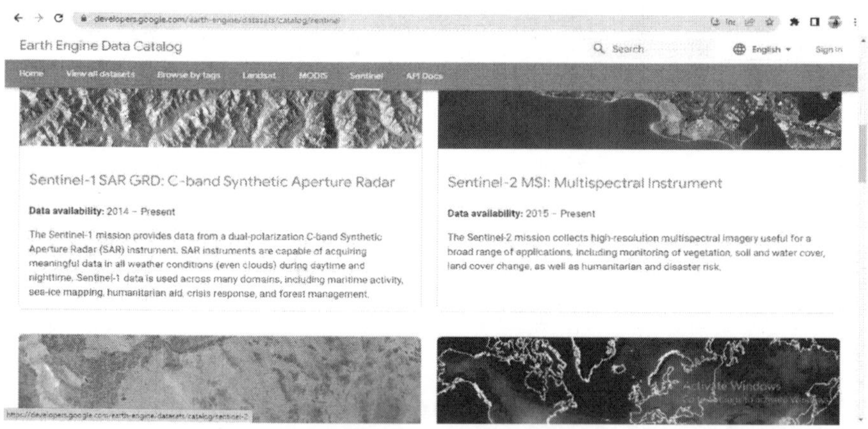

Figure 11.2: Seninel data repository

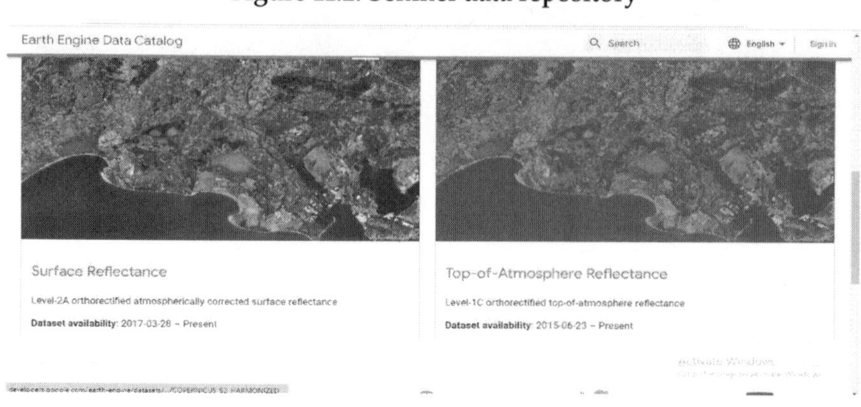

Figure 11.3: Select surface reflectance

On selection of the data, the following code will show. Copy the code and paste in the newscript or select the tab –open in Code Editor given bellow.

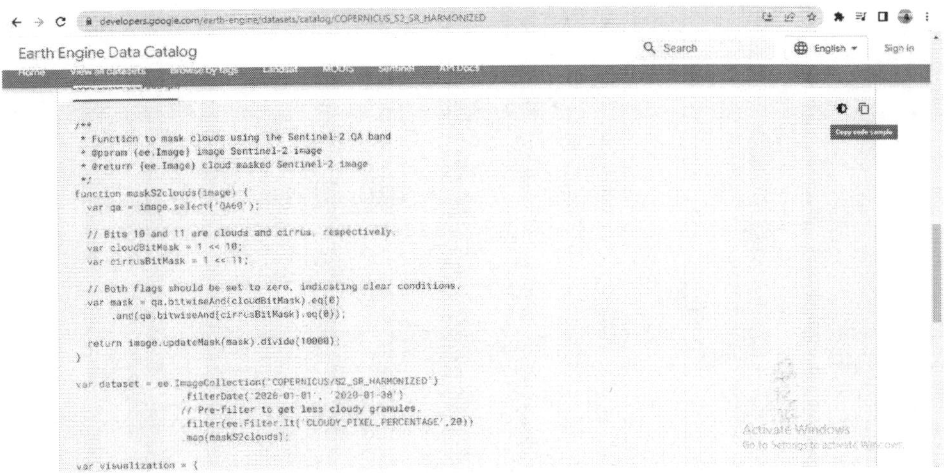

Figure 11.4: Code for the sentinel 2

Open Code Editor

Paste the copied code into a new script

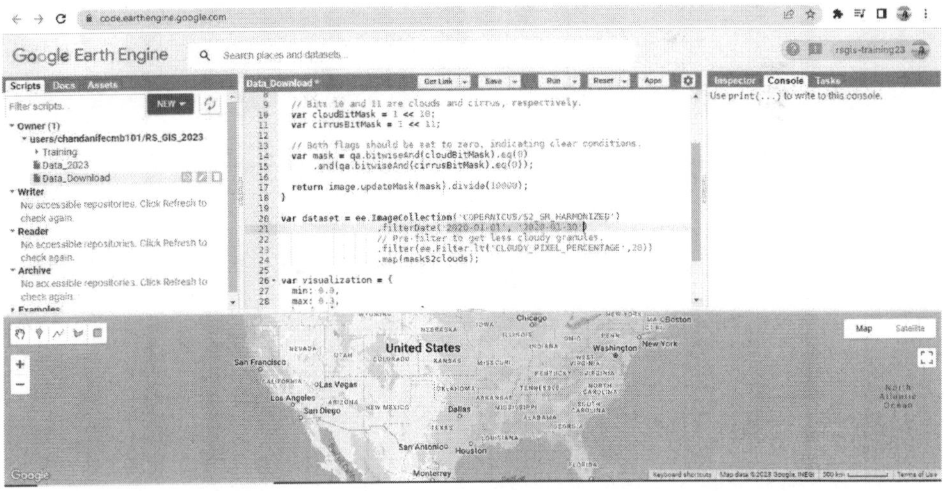

Figure 11.5: Paste the code of sentinel 2 in code editor

- Add the shape file of your study area to the script as shown before.
- Rename the shape file to –aoi

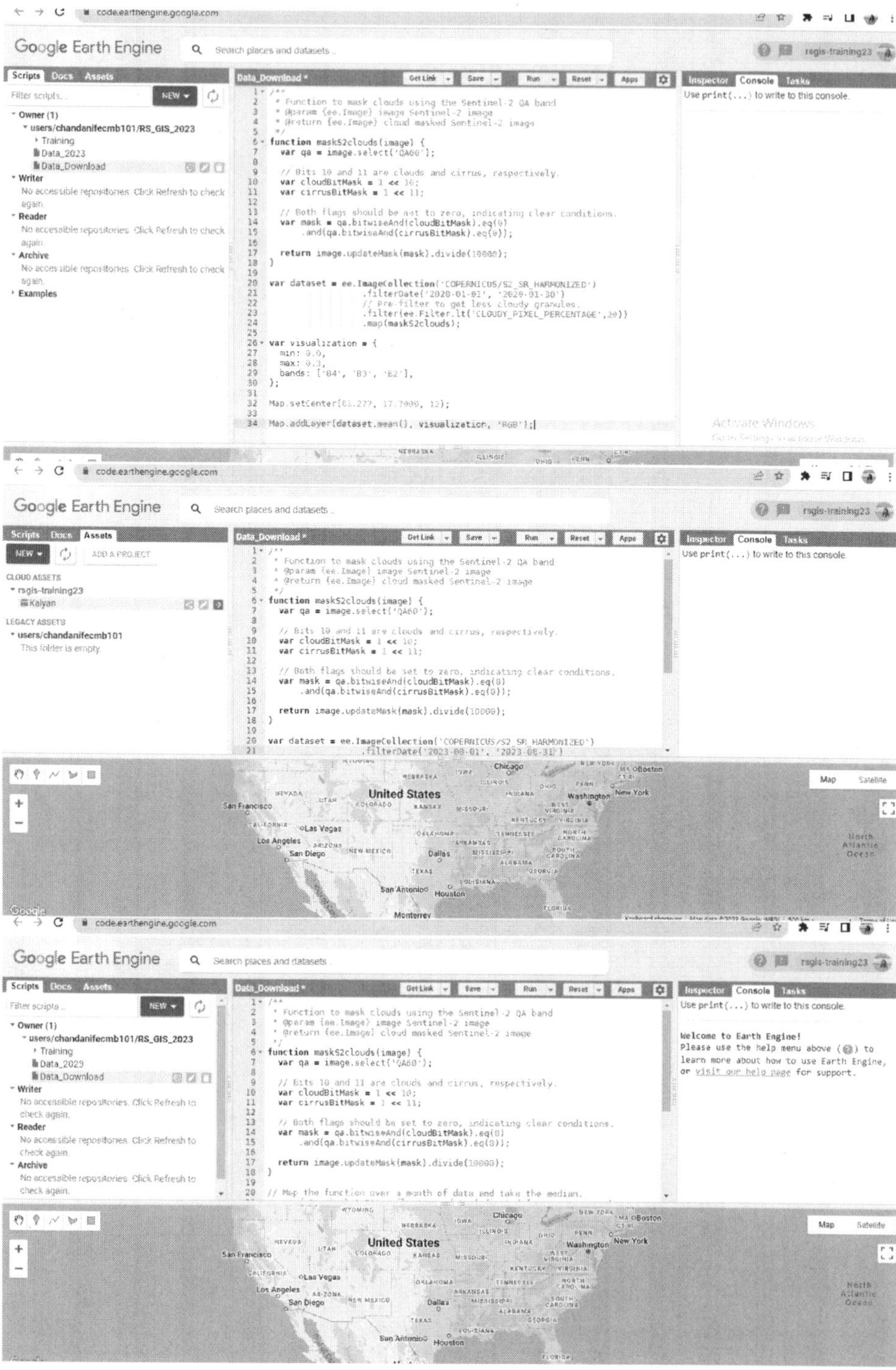

Figure 11.6: Add the shape file and rename it

- Clip the dataset to the shape file using the code var s2clip = s2.clip(aoi);

  Map.addLayer(s2clip, visualization, 'Clipped');
- Export the image to drive using the following code Export.image. toDrive({image:s2clip, description: 'Name of the data', folder: 'Name of the folder', region:aoi, scale:10, maxPixels:10e9, fileFormat:'GEOTIFF' });
- Run the code

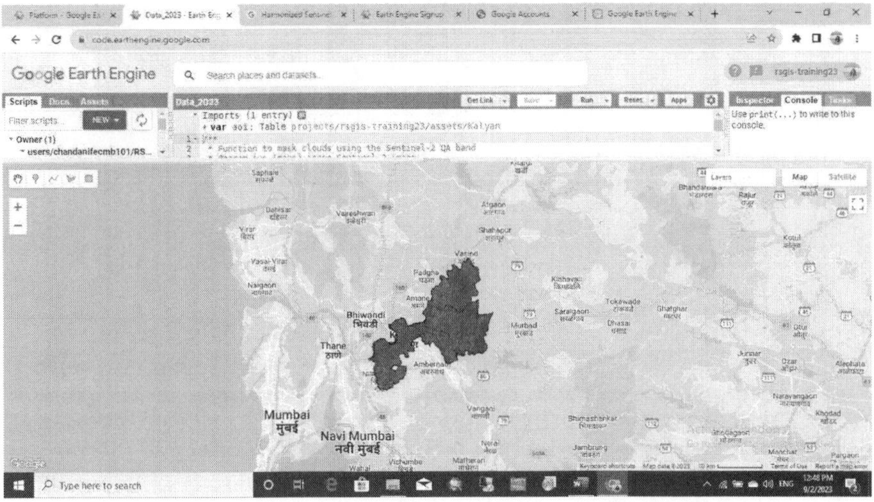

Figure 11.7: Sentinel 2 image clipped to the study area

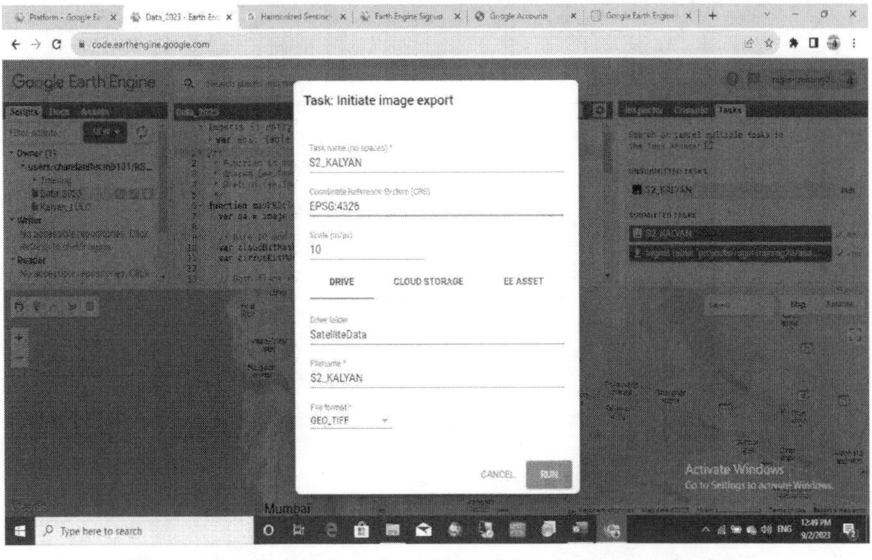

Figure 11.8: Exporting the clipped image to google drive

The resulting image after applying the clipping operation will resemble this. Go to the Task button

Fill in all the details as follow

After successful export open in drive download

CHAPTER 12

LAND USE LAND COVER CLASSIFICATION

Vinod Kumar Yadav[1], Chandani Dave[1] and Mohammed Meharoof[2]

[1] ICAR-Central Institute of fisheries Education, Mumbai-400061
[2] Assistant Professor (o/c), College of Fisheries, Payyanur, Kannur-670307

Introduction

Land Use Land Cover (LULC) classification categorizes Earth's surface into different classes or categories based on the predominant land use and land cover types. GEE provides a powerful platform for conducting LULC classification using satellite imagery and geospatial data. This process enables researchers, environmentalists, urban planners,and policymakers to gain valuable insights into changes in land cover, monitor natural resources, assess urban expansion, and track environmental trends over time. By leveraging GEE's extensive datasets and computational capabilities, users can performaccurate and efficient LULC classification to support various applications, from ecological conservation to land management and urban development planning.

Supervised Classification:

Supervised classification is a machine learning method where a computer algorithm learns to categorize data (e.g., pixels in an image) into predefined classes based on a labeled dataset. It uses known examples to recognize and classify new, unlabeled data accurately.Following are the machine learning models which can be used for supervised classification in GEE.

Random Forest: Random Forest is a popular ensemble learning algorithm for LULC classification. It is robust and can handle both spectral and non-spectral features.

Support Vector Machines (SVM): SVM is effective for binary and multi-class classification tasks and works well with high-dimensional data such as remote sensing imagery.

Classification and Regression Trees (CART): is a decision tree-based machine learning algorithm used for classification and regression tasks, creating a binary treestructure to make predictions based on feature values.

Steps for the classification

Several steps involve performing a LULC classification using the Random Forest algorithmin Google Earth Engine (GEE). Below are the key steps to carry out this task:

1. Data Selection:

Choose satellite imagery or remote sensing data sources suitable for your LULC classification. Common options include Landsat, Sentinel, or other high-resolution imagery.

Code for this step

```
/**
 * Function to mask clouds using the Sentinel-2 QA band
 * @param {ee.Image} image Sentinel-2 image
 * @return {ee.Image} cloud masked Sentinel-2 image
 */
function maskS2clouds(image) {
  var qa = image.select('QA60');

  // Bits 10 and 11 are clouds and cirrus, respectively.
  var cloudBitMask = 1 << 10;
  var cirrusBitMask = 1 << 11;

  // Both flags should be set to zero, indicating clear
  conditions.var mask = qa.bitwiseAnd(cloudBitMask).
  eq(0)
      .and(qa.bitwiseAnd(cirrusBitMask).eq(0));
  return image.updateMask(mask).divide(10000);
}
var dataset = ee.ImageCollection('COPERNICUS/S2_SR_HARMO-
NIZED')
            .filterDate('2020-01-01', '2020-01-30')
            // Pre-filter to get less cloudy granules.
            .filter(ee.Filter.lt('CLOUDY_PIXEL_PERCENTAGE',20))
            .map(maskS2clouds)
            .select(['B8','B4','B3','B2','B11'])
            .filterBounds(aoi);
```

```
var visualization = {
  min: 0.0,
  max: 0.3,
  bands: ['B8', 'B4', 'B3'],
};

Map.centerObject(aoi,20);
var s2= dataset.median();
// Map.addLayer(dataset.median(), visualization, 'RGB');
// Map.addLayer(aoi);
// Map.addLayer(s2, visualization,'RGB');
```

2. Region of Interest (ROI) Selection: Define the area of interest (ROI) for your classification. You can create a polygon feature in GEE to specify the study area.

- Add the shape file of study area as shown in previous exercise

3. Data Preprocessing:

Clip the satellite imagery to the ROI to reduce computation and focus on the study area.

Code for the step

```
var s2clip = s2.clip(aoi);
Map.addLayer(s2clip, visualization, 'Clipped');
```

4. Training Data Collection:

Create a labeled dataset by selecting representative sample points within your ROI. Each point should be associated with a land cover class. You can use the FeatureCollection in GEE to create this training dataset.

Code for the step

5. Feature Extraction:

Extract spectral and other relevant features (e.g., vegetation indices, texture) from the satellite imagery for each sample point in your training dataset.

6. Random Forest Model Setup:

Import the Random Forest classifier from the GEE library using ee.Classifier.randomForest()and configure its parameters (e.g., number of trees, tree depth).

7. Training the Model:

Train the Random Forest classifier using the training dataset and the extracted features. Use the .train() method to do this.

Code for the steps

```
var training = s2clip.sampleRegions({collection:-
            classlulc,properties: ['Label'],
            scale: 10,
});
var ClassifierRF  = ee.Classifier.smileRandomForest(20).train({fea-
tures:training,
             classProperty: 'Label',
            inputProperties: s2clip.bandNames(),
});


var ClassifiedRF = s2clip.classify(ClassifierRF);
Map.addLayer(ClassifiedRF,{min:0,
max:4,
    palette:['218e02','98ff00','093d72','ffc82d','d02505']},'LULCRF');
```

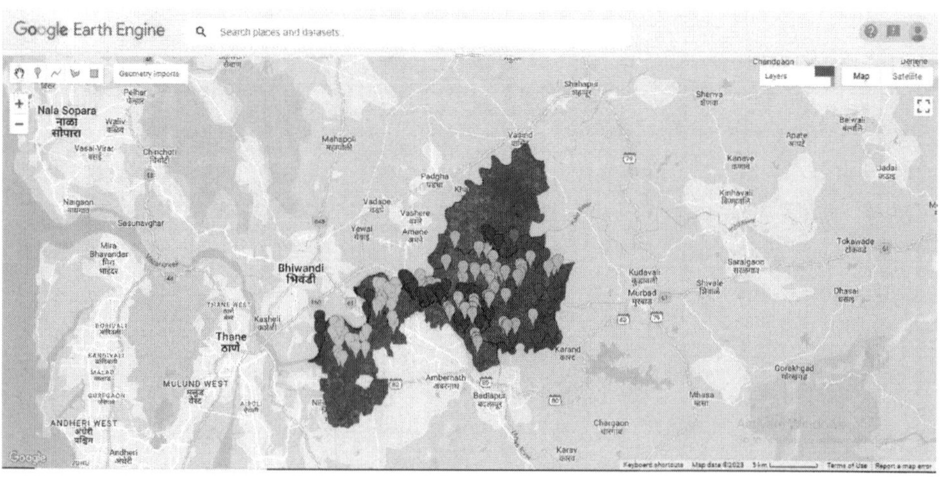

Figure 12.1: Ground Control Points for training data

8. Classification:

Apply the trained Random Forest classifier to the entire image or region of interest toperform the LULC classification. Use the .classify() method on the image data.

Output of Classification

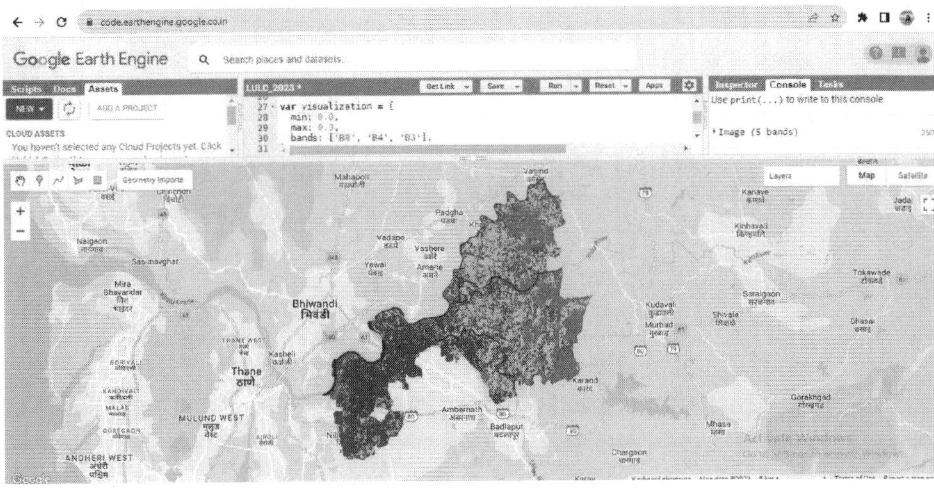

Figure 12.2: Classified image

9. Accuracy Assessment (Optional):

Assess the accuracy of your classification by comparing it to ground truth data or by usingvalidation datasets.

```
var validation = classifiedRF.sampleRegions({collec-
        tion:validationGCP,properties: ['label'],
        scale:30,
        // tileScale:16,
});
var testConfusionMatrix = validation.errorMatrix('la-
bel','classification');print ('Error Matrix RF 2002:',test-
ConfusionMatrix);
print ('Test Accuracy RF 2002:',testConfusionMatrix.accuracy());
var trainingAccuracy = classifierRF.confusionMatrix();
print ('Training Accuracy RF 2002:',trainingAccuracy.accuracy());
print ('confusion Matrix RF 2002:', trainingAccuracy);
```

Results of accuracy assessment

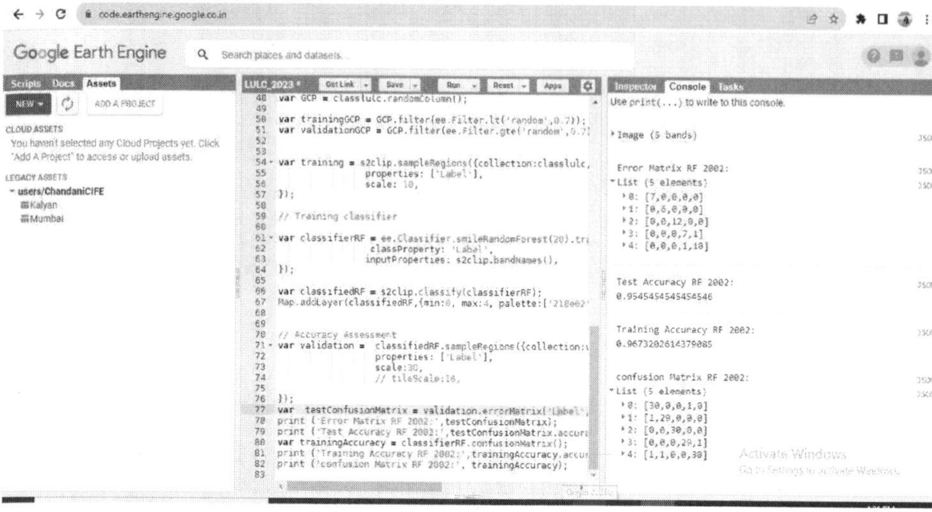

Figure 12.3: Error matrix and confusion matrix

10. Export Results:

If needed, export the LULC classification results as an asset in GEE for further analysis ordownload it for external use.

```
Export.image.toDrive({image:ClassifiedRF,
                     description:'LULC_KALYAN_RF',
                     folder:'SatelliteData',
                     region: aoi, scale:10, maxPixels:10e9,
                     fileFormat:'GEOTIFF'
                     });
```

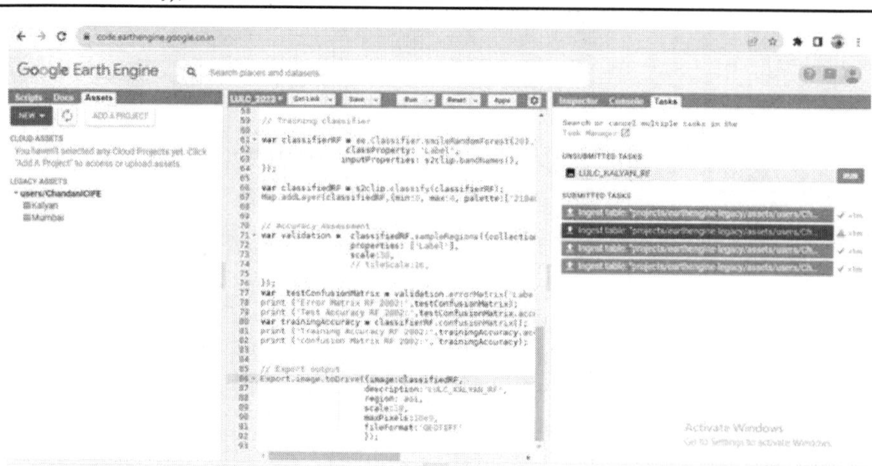

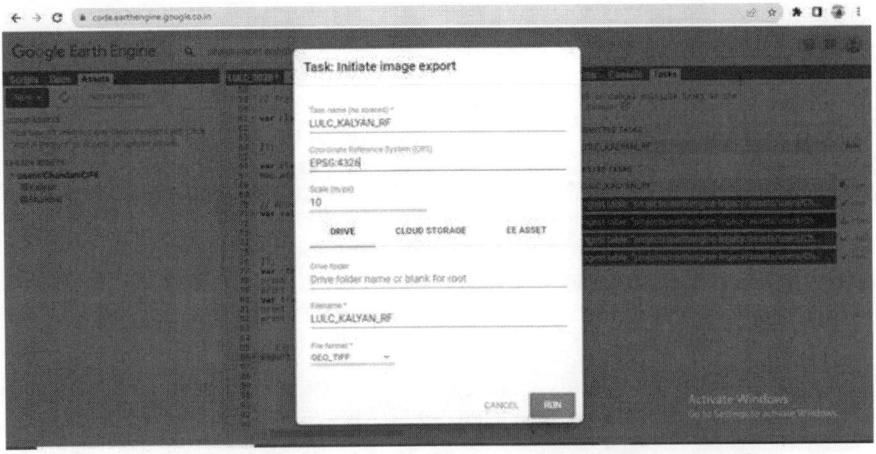

Figure 12.4: export the classified image to google drive

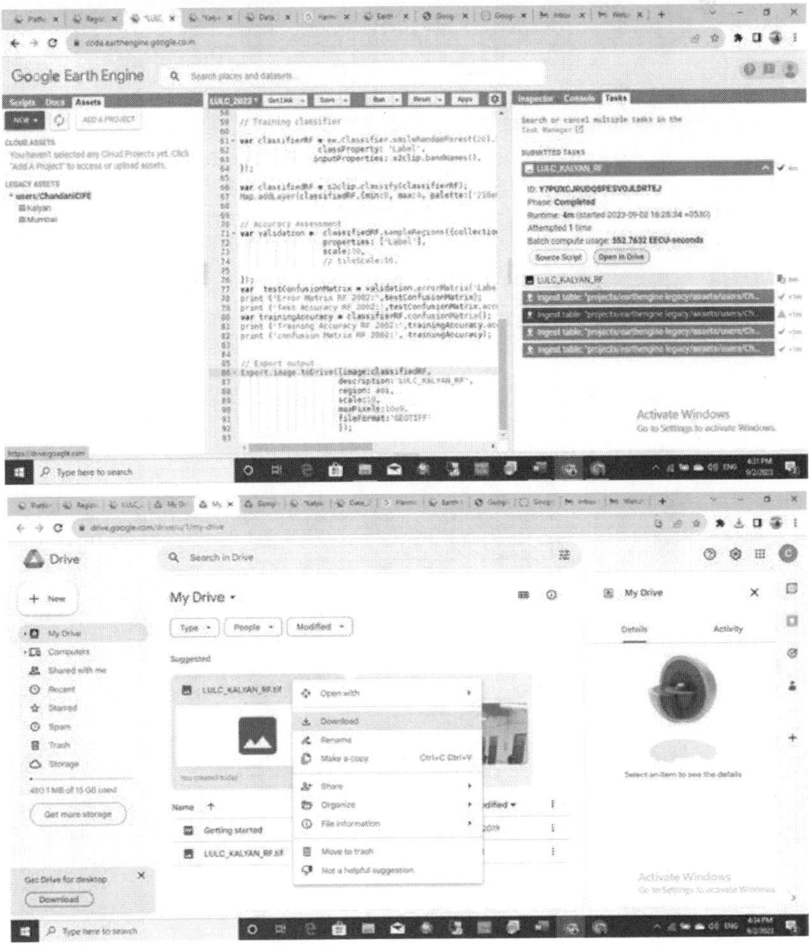

Figure 12.5: Download the classified image from Google drive

Open the output in ArcGIS or QGIS

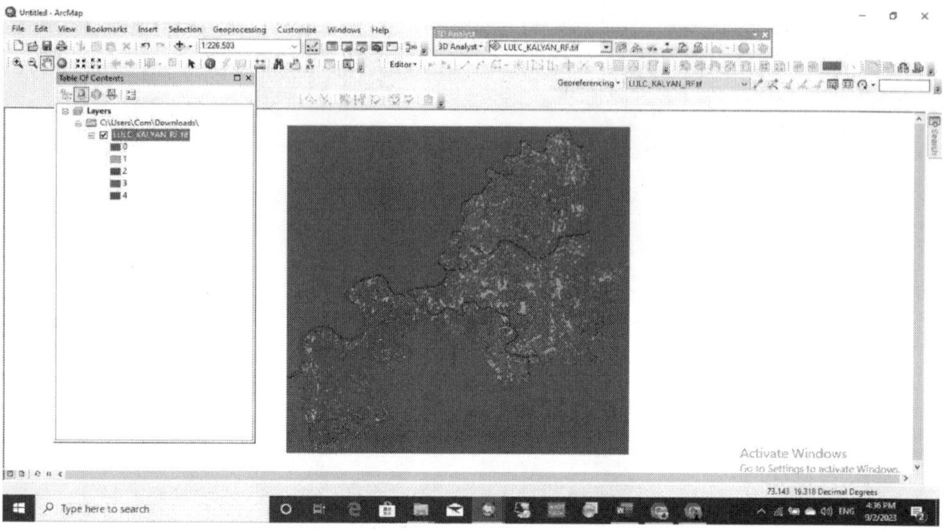

Figure 12.6: Output in the GIS software

Throughout these steps, you'll be using the Earth Engine JavaScript API to execute code within the GEE Code Editor. GEE provides a powerful environment for remote sensing and machine learning tasks, making it well-suited for LULC classification projects.

CHAPTER 13

SPATIAL MAPPING OF RIVERINE HABITAT AND VULNERABLE AREAS FOR FLOW CONNECTIVITY IN RIVERINE ECOSYSTEM FOR CONSERVATION AND MANAGEMENT

Kantarajan G[1] and Vidya Shree Bharti[2]

1 ICAR-National Bureau of Fish Genetic Resources, Lucknow – 226002, Uttar Pradesh

2 ICAR-Central Institute of fisheries Education, Mumbai-400061

The GIS and Remote Sensing tools are extensively used in aquatic environmentmanagement to assess the spatio-temporal water spread dynamics of waterbodies such as rivers, reservoirs, lakes, ponds and coastal waters. The monitoring of surface water is important to understand its ecological and hydrological processes to plan for sustainable management.

Geographical Information System

GIS is defined as "a computer system capable of capturing, storing, analyzing, and displaying geographically referenced information; that is, data identified according to location" by United States Geological Survey. The major componentsof GIS include Hardware (computer for operation), Software (tools to store, analyze, and visualization of spatial information), Data (geographic and tabular data), People (technical specialist) and Methods (plans and defined rules).

Remote sensing

Remote sensing is "the process of detecting and monitoring the physical characteristics ofan area by measuring its reflected and emitted radiation at a distance" (USGS). The following are the essential components of Remote Sensing (Fig. 1): Energy source, Interaction with the target, Recording of energy by the sensor, Transmission, Reception,and Processing and Interpretation and Analysis (Panigrahy and Ray, 2006).

Interaction of Important earth surface components with Electro-MagneticRadiation (EMR)

A portion of EMR which strikes on the earth surface may be reflected, absorbed, transmitted or scattered depends on the wavelength of the radiation. The features/properties of the target can be measured by the incident energy reflected by the surface. The spectral reflectance also varies according to the EMR (Aggarwal, 2004). The spectral reflectanceof various earth surface features is depicted in Fig. 2.

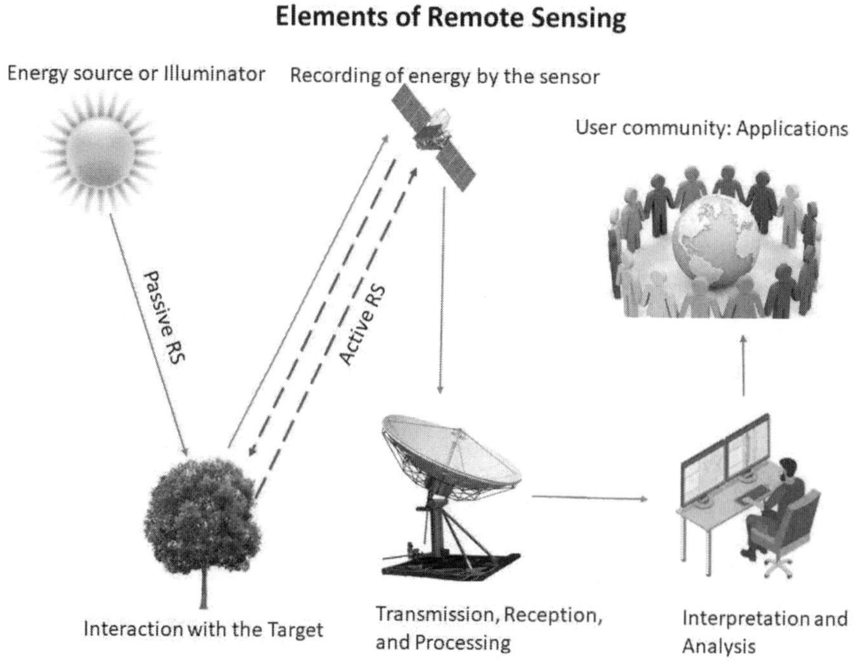

Figure 13.1: The essential components of Remote Sensing

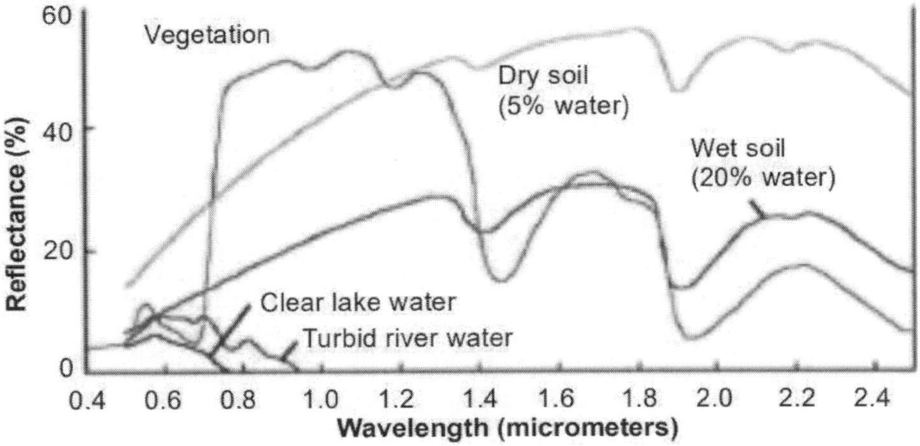

Figure 13.2: Spectral reflectance signatures of soil, water and vegetation (source: Aggarwal, 2004)

Soil: Majority of the incident wave energy is reflected and very little amount, only selected portion is diffused through the soil. The level of reflectance in soil surface slowly increases with wavelength in visible and near infrared region of the spectrum. Moisture in soil reduce its reflectance.

Vegetation: The leaves of the vegetation strongly reflects green radiation and absorbs red and blue wavelengths. The health status of the vegetation may be assessed using near infrared radiation as it acts as a diffuse reflector for this wavelength.

Water: Generally, water reflects shorter wavelength radiation while longer visible wavelength and NIR is absorbed. Hence, pure water has high reflectance in Visible wavelength (0.4 and 0.6 μm) whereas no reflectance is reported in NIR and higher wave length radiation. So clear water appears dark in higher wave lengths. The reflectance of a water body is mainly influenced by depth, suspended solids in water and roughness of water surface. Estuarine water with high suspended solids reflects more at longer wavelengths of the visible spectrum (Yellow/Red range).

Applications of satellite datasets in Aquatic Environment Management

The satellite datasets which are mainly used in aquatic ecosystem management studies are grouped in to 3 different classes, as follows (Huang et al., 2018);

i. Coarse resolution (pixel size: >200 m): These remote sensors offer low accuracy spatial resolution and high temporal resolution with broad coverage. The MODIS-Moderate Resolution Imaging Spectroradiometer of NASA - Terra and Aqua satellites is a very popular among this group it offers 250 m (visible)

- 500 m (infrared band) dataset and daily revisit times across different regions of the globe. These datasets have been widely used for estimating the surface water spread and near-real-time flood monitoring programs.

ii. Medium resolution (pixel size: 5-200 m): These sensor satellite has finer resolution than the coarse resolution satellite but the temporal resolution and swath area is very less. The major sensor types which are predominantly used among the **Indian Remote SensingSatellites** Datasets i.e. IRS 1A and 1B: LISS-I (73 m) and LISS-II (36.25 m); IRS 1C: PAN: <6 m; LISS III: 23.6 m; IRS 1D: PAN (6 m), LISS III (23.6 m); IRS P6 / Resourcesat 1:

LISS-III (23.5 m), AWiFS and a High-Resolution Multispectral Sensor LISS-IV (5.8 m at nadir) (https://www.nrsc.gov.in/; https://www.isro.gov.in/spacecraft/list-of-earth- observation-satellites).

USGS/NASA: Landsat Data

The Landsat, the longest continuous space-based record of Earth's land since 1972, under the series of Earth-observing satellites co-managed by USGS & NASA (https://www.usgs.gov/science-explorer-results?es=Landsat). These datasets have been extensively used for LULC mapping, resource inventory, mapping and monitoring purposes. The repetitive and synoptic coverage of Landsat series dataset enabled the LULC change analysis over the time period. The sensor characteristics of Landsat satellite (Miller, 2016) is given in Table 1.

Table 1. Comparison of features of selected Landsat sensors

MSI Characteristics	MSS (Landsat 1-5)	TM (Landsat 4,5)	ETM+ (Landsat 7)	OLI & TRS (Landsat 8)
Spatial resol. (m)	90	30	30	30
No. of bands	4	6	7	9
Bit depth	8-bit	8-bit	8-bit	12-bit
File size	25 MB	170 MB	250 MB	1 GB

Note: MSS, Multispectral Scanner; TM, Thematic Mapper; ETM+, Enhanced Thematic Mapper Plus; OLI, Operational Land Imager; TIRS, Thermal Infrared Sensor

ESA: Sentinel Data

The Sentinel family satellites were managed by European Space Agency (ESA) which are mainly used to observe the land, ocean and atmosphere.

Sentinel-1 using synthetic aperture radar (SAR) with 9-40 m spatial resolution uses the microwave pulses for imaging. These datasets are mainly used in flood mitigation programmes (https://gisgeography.com/sentinel-satellites-copernicus-programme/). This data is highly preferred for continuous monitoring of surface waterbody as it has the mapping ability in all weather conditions.

Sentinel 2 mission's primary objective is to provide high resolution satellite data for land cover/use monitoring, climate change and disaster monitoring. (https://www.esa.int/Applications/Observing_the_Earth/Copernicus/The_Sentinel_missions)

. The spatial resolution and band composition of Sentinel 2A data is given in Table 2. The revisit time period of 2A and 2B satellite is 5 days (ESA, 2021). The combination of high resolution (10 m), novel spectral capabilities, a swath width of 290 km and frequent revisit time (5 days) and free access policy leads to high acceptance and application of Sentinel 2 images.

Table 2. Band composition and characteristics of Sentinel 2A imagery

Band	Spectral region	Wavelength (nm)	Resolution (m)
B1			
B2	Blue	458-523	10
B3	Green peak	543-578	10
B4	Red	650-680	10
B5	Red edge	698-713	20
B6	Red edge	733-748	20
B7	Red edge	773-793	20
B8	NIR	785-899	10
B8A	NIR narrow	855-875	20
B11	SWIR	1565-1655	20
B12	SWIR	2100-2280	20

iii. High resolution (pixel size: <5 m) sensors are suitable for detecting small features. Sensors such as IKONOS, RapidEye, Worldview, ZY-3, Quickbird, and GF-1/2 provides images with meter – sub-meter level resolution. However, these sensors have several limitations: small scene coverage – unsuitable for mapping large waterbodies; appearance of shadow – affects the water detection; limited image availability and revisit time period of these sensors is uncertain. The imageries of sensor enable the detection of waterbodies with high accuracy, but due to its lower temporal resolution. Cost and limited availability of spectral bands not suitable for intensive temporal monitoring studies.

Extraction and mapping of surface water: Spectral Indices

The spectral indices are most commonly used to identify and map the water areas from other land use categories. The most commonly used indices are as follows;

i. Normalized Difference Vegetation Index - NDVI (Rouse et al., 1974)

Where, Red is the spectral reflectance measurement of red band and NIR is the spectral reflectance measurement of near-infrared band. The NDVI values varies between -1.0 to +1.0. The NDVI values of features such as water, snow and cloud ranges between -1.0 to 0 while for barren land it falls within the range of 0-0.2. For vegetation the value falls between 0.2-1.

ii. Normalized Difference Water Index - NDWI (McFeeters, 1996)

It is considered as a stable water index, and is suitable for identifying water pools in regions under different hydro-climatological influence.

Where, Green is the spectral reflectance measurement of green band and NIR is the spectral reflectance measurement of near-infrared band. The NDWI value ranges from -1.0 to +1.0. The positive NDWI values indicates water features while the range of values between -1 to 0indicates vegetation and soil features (Fig. 3).

The NDWI separating water body and vegetation satisfactorily but its application is limited in soil and built-up area (Gautam et al., 2015). Seaton et al. (2020) demonstrated that the NDWI computed from Sentinel-2 TOA reflectance imagery has shown higher accuracies over other indices viz., MNDWI, NDVI, AWEIsh, AWEInsh, MBWI in mapping the pools along the perennial and non-perennial rivers in different hydro-climatological regions of the Western Cape using Sentinel and Landsat imageries. NDWI was most accurate for surface water mapping when using Landsat or Sentinel (Seaton and Dube, 2021).

iii. Modified Normalized Difference Water Index – MNDWI (Xu, 2006)

The MNDWI is capable of reducing the built-up area noise hence it is highly suitable for extracting water information for a water region with a background dominated by built-up land.

Where, Green is the spectral reflectance measurement of green band and SWIR is the spectral reflectance measurement of shortwave infrared band. The MNDWI value ranges from -1.0 to +1.0. However, compared to NDWI, water bodies have greater +ve values and soil, vegetation and built-up classes have smaller -ve values in MNDWI. MNDWI is considered as the most efficient spectral index in enhancing water related features as the values of water features never mixed with other land cover type.

Limitations of Sentinel-2 data based surface water mapping

The notable limitations of using Sentinel 2-MSI data for river mapping Seaton et al. (2020) islisted below;

iii. Classification was greatly affected by cloud and mountain shadows; however, atmospheric correction seemed to minimise the misclassification of shadows, but at expense of eliminating many true water pixels.

ii. The effects of riparian vegetation on the classification of surface water can be minimised by use of in-situ measurements to improve mapping.

iii. The spatial resolution of Sentinel-2 limited the study to mapping pools with surface areas greater than 100 m^2, with smaller pools being undetectable.

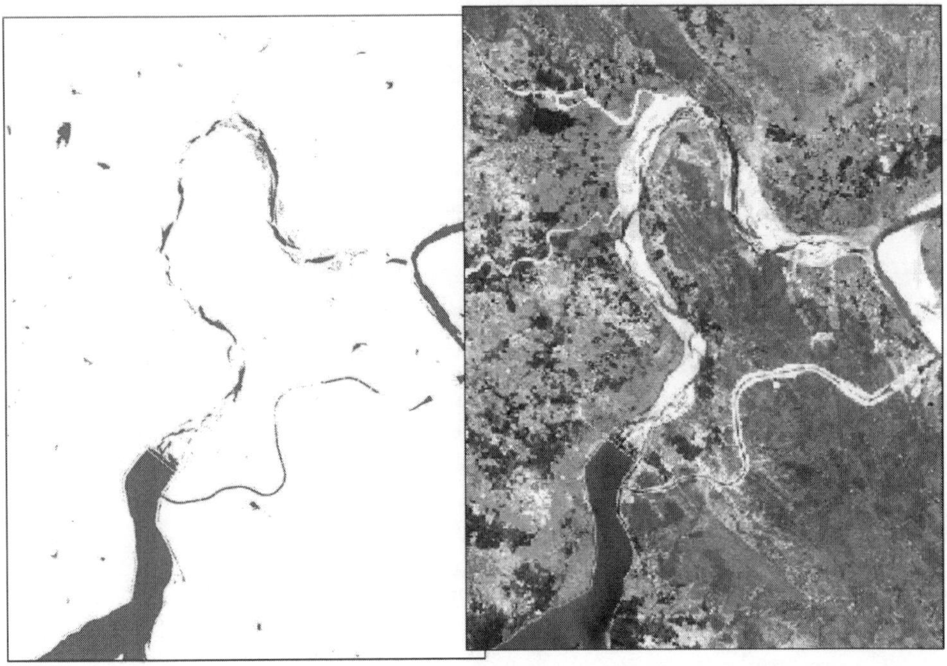

Figure 13.3: Sentinel based NDWI map depicting the water pixels along the Godavari river

Assessment of temporal water spread frequency and mapping of vulnerable areas

The continuous presence of water along the river stretch can be assessed by integrating the reclassified NDWI images of different periods, *viz.*, month, season, and year using the WaterPresence Frequency (WPF) method (Borro et al., 2014; Das & Pal, 2016). Likewise, perennial riverine habitats i.e., deep pools that hold water even during the dry season can be mapped using the Water Presence Frequency approach. The WPF was estimated for generating the temporal composite image of water presence.

Where WPFj, water presence frequency of jth pixels in a time period;Ij, jth pixel having water in the selected NDWI images;

n, number of images.

The values of WPF image pixels range from 0 to 100%. A higher value indicates the frequent/regular presence of water in the pixel, and a lower value indicates the irregularity in the water presence. Pixel with 100 indicates the presence of water during the entire period, and zero value denotes _no water".

Stream continuity can be assessed by visualising the water presence composite imageof a particular period. The pixels which had 100% WPF aggregated and converted into a vector shapefile. The critical manual evaluation of the polygons for their continuity, and possible breakages between them can be spatially delineated to identify the vulnerable areasfor riverine discontinuity.

Steps to calculate NDWI and extraction of water class in QGIS

Extraction/Calculation of WSA:

Download data

Import river boundary: How to delineate?

Import required bands to QGIS

Built virtual raster

Symbology → Choose Multi band color → Arrange bandsMerge tiles (to make a single file for your study area Calculate NDWI (Green - NIR) / (Green + NIR))

Raster → Raster calculator → Type formula expressions

Symbology → Select single point Pseudo color → Interpolation: Discrete → Mode: EqualInterval → Classify in to different classess

Reclassify by Table: 0-1 → as water class

Raster layer Unique values report – area calculation

Processing Toolbox → Raster Analysis → Raster layer Unique values

Assessment of WPF:

Import Water class imagery of a specific period of analysisRaster → Raster calculator → Type formula expressions i.e. ((sum of imageries)/ Total no. of imageries) x 100

Classification: High / Medium / Low

Classification of permanent water pixels: pixels with 100% WPF View of working window:

Figure 13.4: Working window of Qgis

Import satellite imagery/bands

Layer → Add layer → Add raster layer

Figure 13.5: Import satellite bands

Calculate NDWI (Green - NIR) / (Green + NIR)

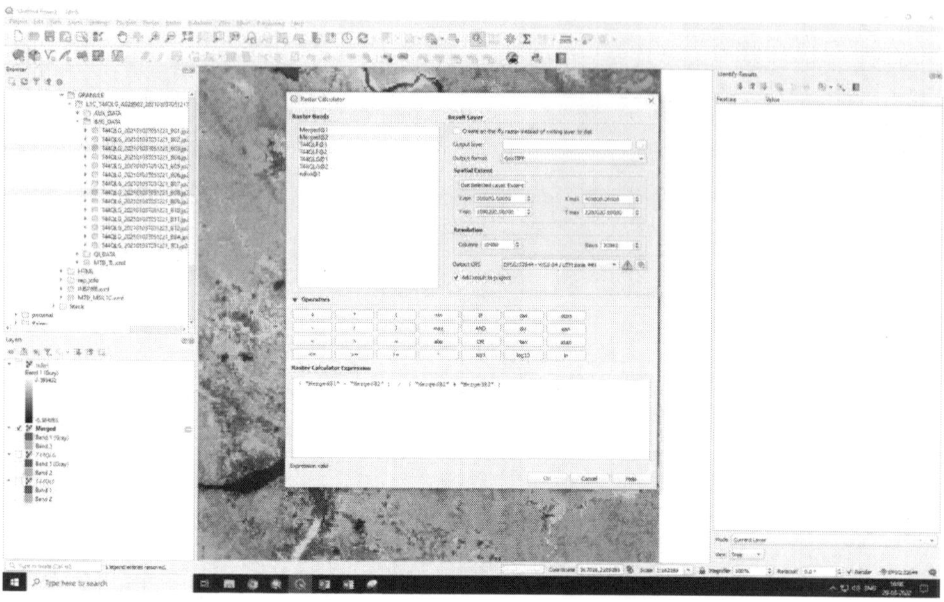

Figure 13.6: Raster calculator

Raster → Raster calculator → Type formula expressions

Extraction

Raster → Extraction → Clip raster by extent

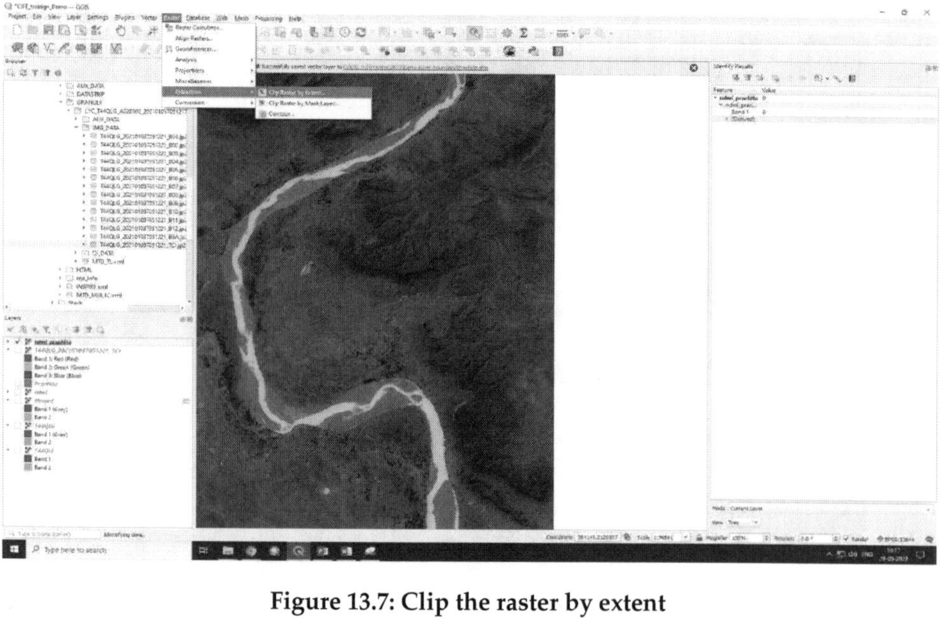

Figure 13.7: Clip the raster by extent

Extraction

Raster → Extraction → Clip raster by mask

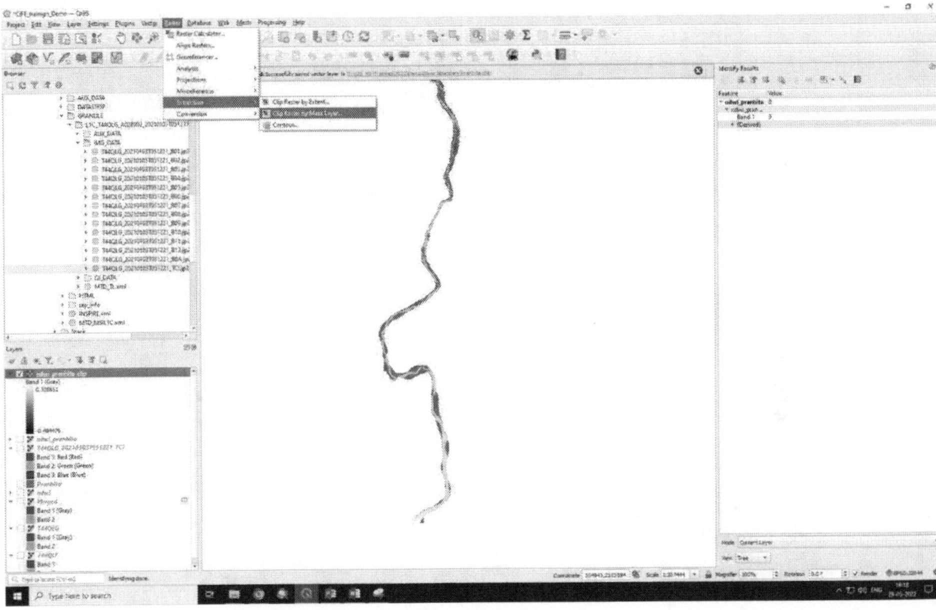

Figure 13.8: Clip the raster by mask

Reclassify by Table: 0-1 → as waterclass

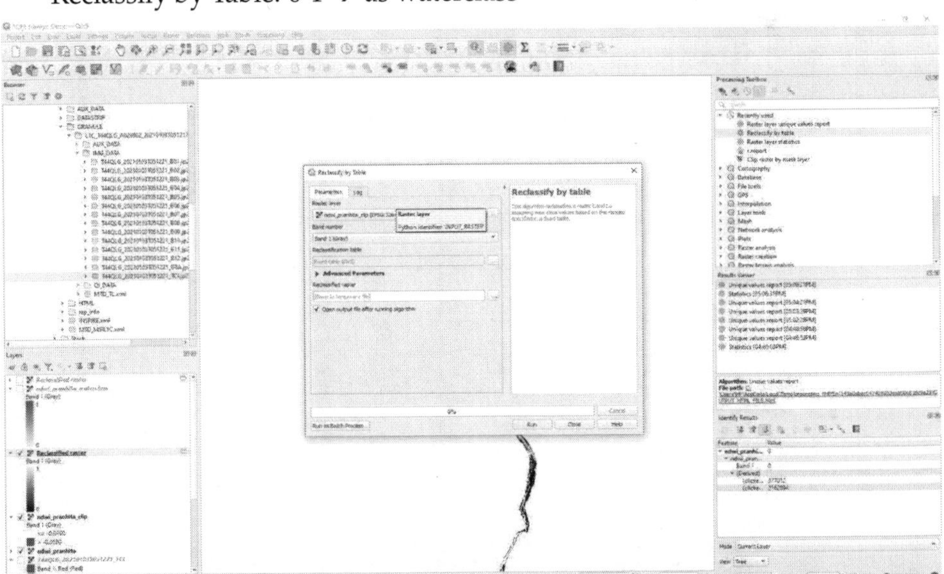

Figure 13.9: Reclassify the data

Raster layer Unique values report – area calculation

Processing Toolbox → Raster Analysis → Raster layer Unique values

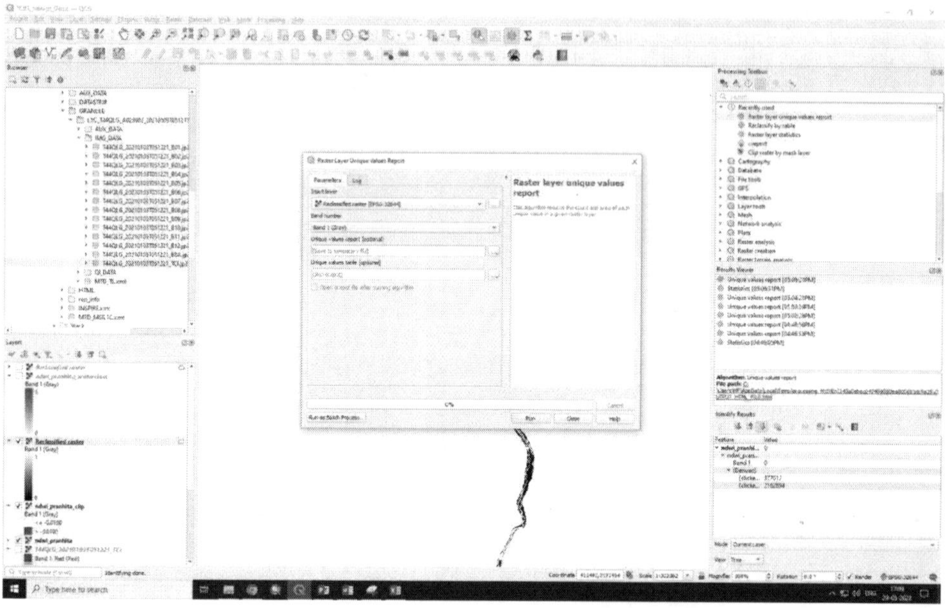

Figure 13.10: Raster layer unique values report

Convert Raster to Vector

Raster → Conversion → Polygonize

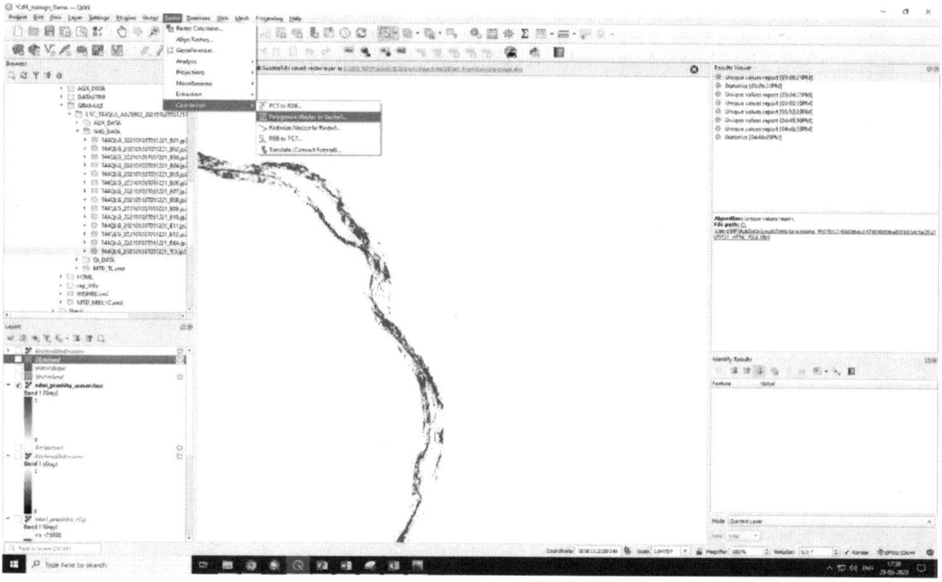

Figure 13.11: Poligonize

References

Aggarwal, S. (2004). Principles of remote sensing. Satellite remote sensing and GISapplications In agricultural meteorology, 23, 23-28.

Borro, M., Morandeira, N., Salvia, M., Minotti, P., Perna, P., & Kandus, P. (2014). Mapping shallow lakes in a large South American floodplain: a frequency approach on multitemporal Landsat TM/ETM data. Journal of Hydrology, 512, 39-52.

Das, R. T., & Pal, S. (2016). Identification of water bodies frommultispectral Landsat imageries of Barind Tract of West Bengal.International Journal of Innovative Research and Review, 4,26–37.

ESA, 2021. Introducing Sentinel-2. The European Space Agency. Available:https://www.esa.int/Applications/Observing_the_Earth/Copernicus/Sentinel- 2/Introducing_Sentinel-2 (accessed on 19 Oct. 21).

Gautam, V. K., Gaurav, P. K., Murugan, P., & Annadurai, M. J. A. P. (2015). Assessment of surface water Dynamics in Bangalore using WRI, NDWI, MNDWI, supervised classification and KT transformation. Aquatic Procedia, 4, 739-746.

Gautam, V. K., Gaurav, P. K., Murugan, P., & Annadurai, M. J. A. P. (2015). Assessment of surface water Dynamics in Bangalore using WRI, NDWI, MNDWI, supervised classification and KT transformation. Aquatic Procedia, 4, 739-746.

Huang, C., Chen, Y., Zhang, S., & Wu, J. (2018). Detecting, extracting, and monitoring surface water from space using optical sensors: A review. Reviews of Geophysics,56,333–360

McFeeters, S. K. (1996). The use of the Normalized Difference Water Index (NDWI) in the delineation of open water features. International journal of remote sensing, 17(7), 1425-1432.

Miller, H.M., 2016, Users and uses of Landsat 8 satellite imagery—2014 survey results: U.S.Geological Survey Open-File Report 2016–1032, 27p.

Panigrahy, S and Ray, S.S., 2006. Remote Sensing. In Environment and Agriculture' 9Ed.

K.L. Chadha, M.S. Swaminathan). Malhotra Publishing House, New Delhi.

Rouse, J. W., Haas, R. H., Schell, J. A., & Deering, D. W. (1974). Monitoring vegetation systems in the Great Plains with ERTS. NASA special publication, 351(1974), 309.

Seaton, D., & Dube, T. (2021). A new modified spatial approach for monitoring non- perennial river water availability using remote sensing in the Tankwa Karoo, WesternCape, South Africa. Water SA, 47(3), 338-346.

Seaton, Dn; Dube, T; Mazvimavi, D (2020). Use of multi-temporal satellite data for monitoring pool surface areas occurring in non-perennial rivers in semi-arid environments of the Western Cape, South Africa. ISPRS Journal of Photogrammetry and Remote Sensing, 167(), 375–384.\

Xu, H. (2006). Modification of normalised difference water index (NDWI) to enhance open water features in remotely sensed imagery. International journal of remote sensing, 27(14), 3025-3033.

Weblinks

https://www.nrsc.gov.in/

https://www.isro.gov.in/spacecraft/list-of-earth-observation-satellites

https://www.usgs.gov/science-explorer-results?es=Landsat

https://gisgeography.com/sentinel-satellites-copernicus-programme/

https://www.esa.int/Applications/Observing_the_Earth/Copernicus/The_Sentinel_missions

CHAPTER 14

HANDS ON MANGROVE MAPPING USING OPEN-SOURCE SOFTWARE: QGIS PLATFORM

Ajay D Nakhawa[1], and Vinod Kumar Yadav[2]

1 Mumbai RS of ICAR-CMFRI
2 ICAR-CIFE

This chapter describes the classification of images acquired by Sentinel-2 Satellite. In addition,some of the SCP tools are illustrated.

We are going to classify the following land cover classes:

1. Mangrove
2. Built-up;
3. Non-Mangrove Vegetation;
4. Water.
5. Mud Flats
6. Barren Land

1. Data Download

We are going to **download a Sentinel-2 image** provided by the Copernicus Scientific Data Hub. In particular we are going to use the following Sentinel-2 bands.

- Band 2 - Blue;
- Band 3 - Green;
- Band 4 - Red;
- Band 5 - Vegetation Red Edge;
- Band 6 - Vegetation Red Edge;
- Band 7 - Vegetation Red Edge;
- Band 8 - NIR;
- Band 8A - Vegetation Red Edge;
- Band 11 - SWIR;
- Band 12 - SWIR;

Start a new QGIS project. Open the tab Download images clicking the button in the SCP menu, or the SCP Tools, or the SCP dock. Select the tab Sentinel-2 download.

In Login Sentinels enter the user name and password for accessing data (free registration is required).

In Search area enter:

- UL X (Lon): 72
- UL Y (Lat): 19
- LR X (Lon): 73
- LR Y (Lat): 18

In Search set:

- Date from: 2016-05-06
- to: 2016-05-06

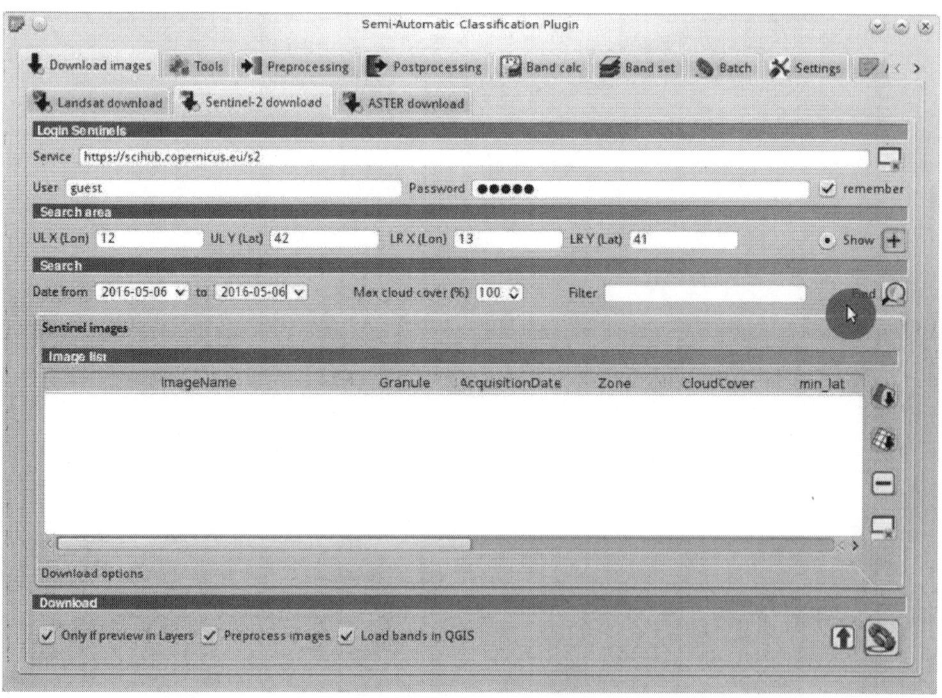

Figure 14.1: Download sentinel 2 data in QGIS project

Search Sentinel-2 images

Now click the button Find  and after a few seconds the image will be listed inthe Image list.

Automatic Conversion to Surface Reflectance: Conversion to reflectance (see Radiance and Reflectance) can be performed automatically. The metadata file (a .xml file whose namecontains MTD_SAFL1C) downloaded with the images contains the required information for the conversion. Read Image conversion to reflectance for informationabout the Top Of Atmosphere (TOA) Reflectance and Surface Reflectance.

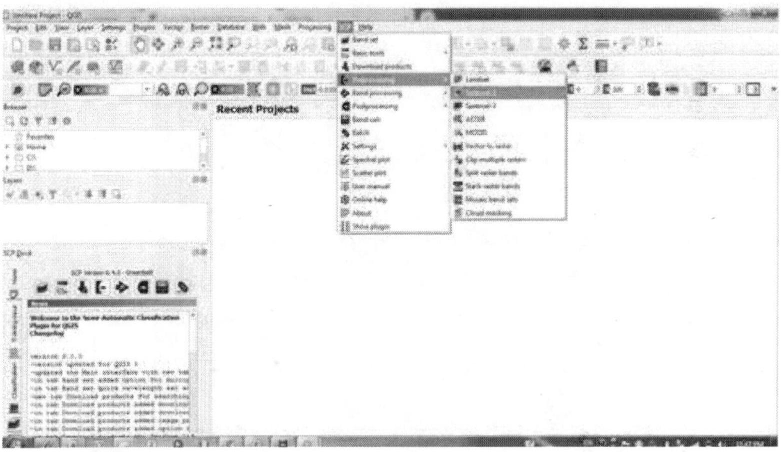

Figure 14.2: Automatic conversion to surface reflection

Sentinel-2 conversion to reflectance

After a few minutes, converted bands are loaded and displayed (file name starts with RT_).If Play sound when finished is checked in Classification process settings, a sound is played when the process is finished.

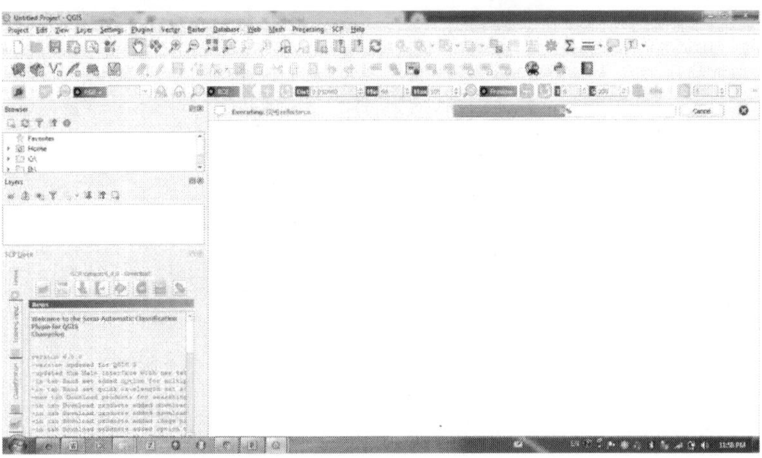

Figure 14.3: conversion to reflectance

2. Clip the Data

Sentinel-2 images have a large extent. In order to reduce the computational time, we aregoing to clip bands to the same study area as Tutorial 1: Your First Land Cover Classification. Open the tab Preprocessing and select the tab Clip multiple rasters.

Click the button ↻ to refresh the layer list, and check all the layers whose name startswith RT_ (the band number is at the end of the layer name).

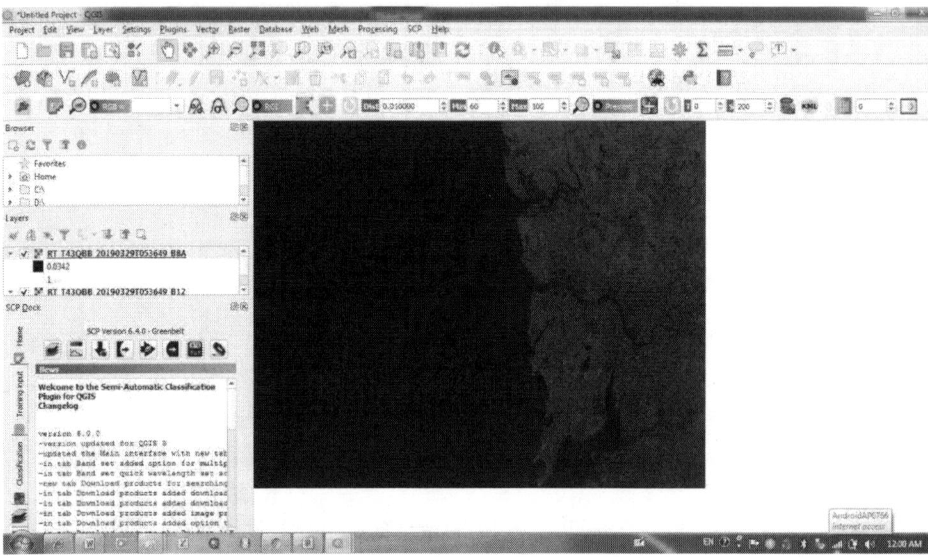

Figure 14.4: Clip the data

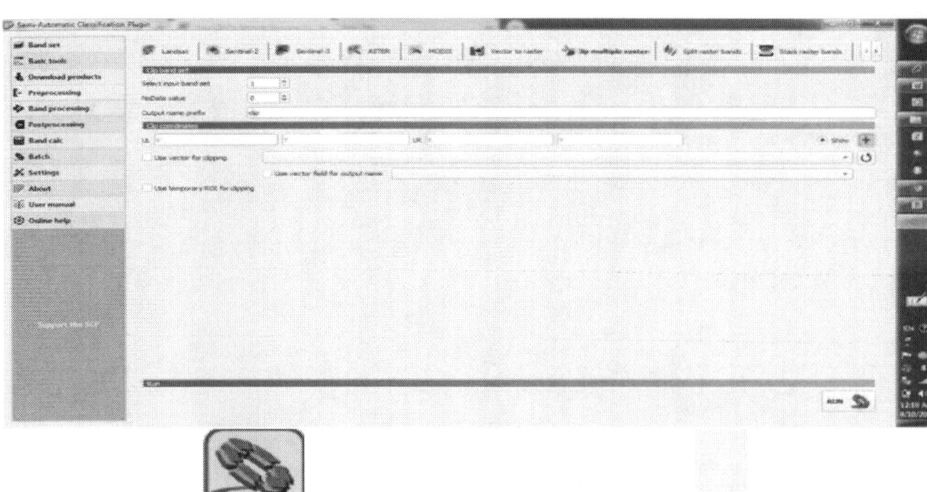

Figure 14.5: Clip the area

- Click the button [+] and select an area such as the following image, or enter the followingvalues:
- UL X: 280034
- UL Y: 2120029
- LR X: 295378
- LR Y: 2098515

clip area: Click the button and select a directory (e.g. clip) where clipped bands aresaved (with the file name prefix defined in Output name prefix). When the process iscompleted, clipped rasters are loaded and displayed. We can remove the bands whose names start with RT_ from QGIS layers.

3. Preprocess the data

The downloaded product is a Sentinel-2 Level 2A image, which is already atmospherically corrected (Level 1C images require atmospherical correction as described in Image conversionto reflectance). Nevertheless, preprocessing is required to convert pixel values to a decimal value of reflectance.

Conversion to reflectance (see Radiance and Reflectance) can be performed automatically. The metadata file (a .txt file whose name contains MTL) downloaded with the images containsthe required information for the conversion.

In order to convert bands to reflectance, open the Preprocessing clicking the button inthe SCP menu or the SCP dock, and select the tab Sentinel-2.

Click the button Directory containing Sentinel-2 bands and select the directory of clippedSentinel-2 bands. The list of bands is automatically loaded in the table Metadata.

You can ignore Select metadata file because this conversion doesn't require additional information.

Sentinel-2 Level 2A images don't require the DOS1 Correction; if it was a Level 1C image weshould enable the option ☑ Apply DOS1 atmospheric correction.

TIP: In general, it is recommended to perform the DOS1 atmospheric correction for the entireimage (before clipping the image) in order to improve the calculation of parameters based onthe image.

For the purpose of this tutorial, uncheck the option ☑ Create Band set and use Band settools because we are going to define this in the following step Define the Band set and createthe Training Input File.

In order to start the conversion process, click the button RUN and select the directorywhere converted bands are saved.

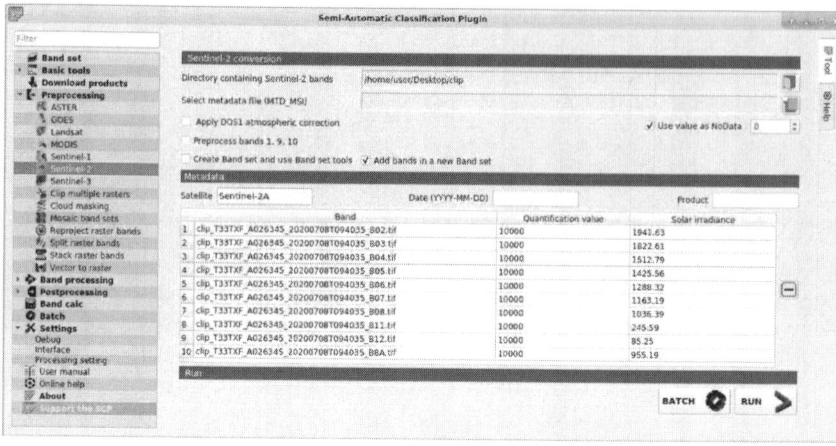

Figure 14.6: Conversion to reflectance

Sentinel-2 conversion to reflectance

After a few minutes, converted bands are loaded and displayed (file name beginning with RT_). If ☑ Play sound when finished is checked in Classification process settings, asound is played when the process is finished.

We can remove all the bands loaded in QGIS layers except the ones whose name beginwith RT_.

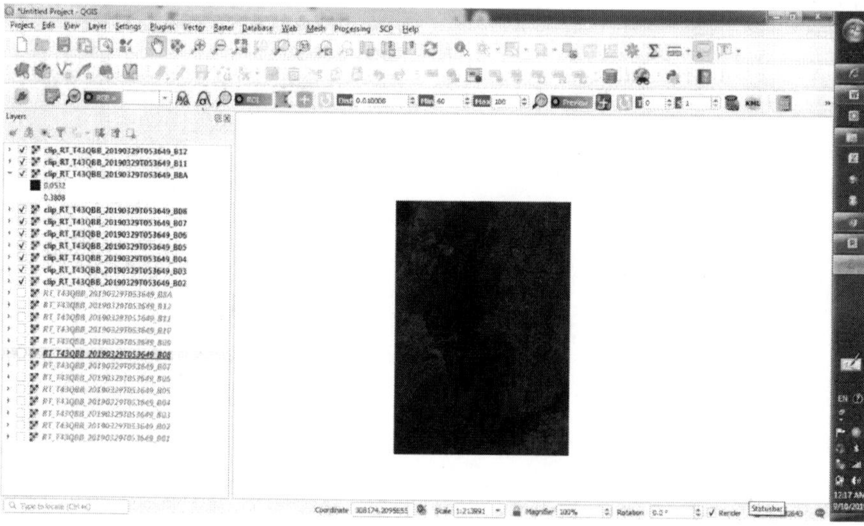

Figure 14.7: Remove extra bands

4. Create the Band Set

Now we need to define the Band set which is the input image for SCP. Open the tab Bandset clicking the button in the SCP menu, or the SCP Tools, or the SCP dock.

Click the button to refresh the layer list, and check all the clipped bands; then click to add selected rasters to the Band set. In the table Band set definition order the band names in ascending order (click to sort bands by name automatically), then highlight band (i.e. single click on band name in the table) and use the buttons or to place this band at number 8. Finally, select Sentinel-2 from the list Quick wavelength settings, in order to set automatically the Center wavelength of each band and the Wavelength unit (required forspectral signature calculation).

5. Create the ROIs

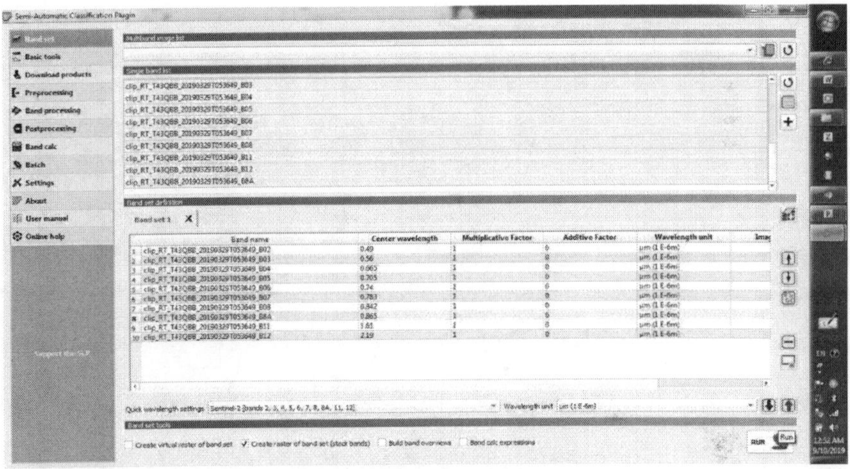

Figure 14.8: Create the ROI

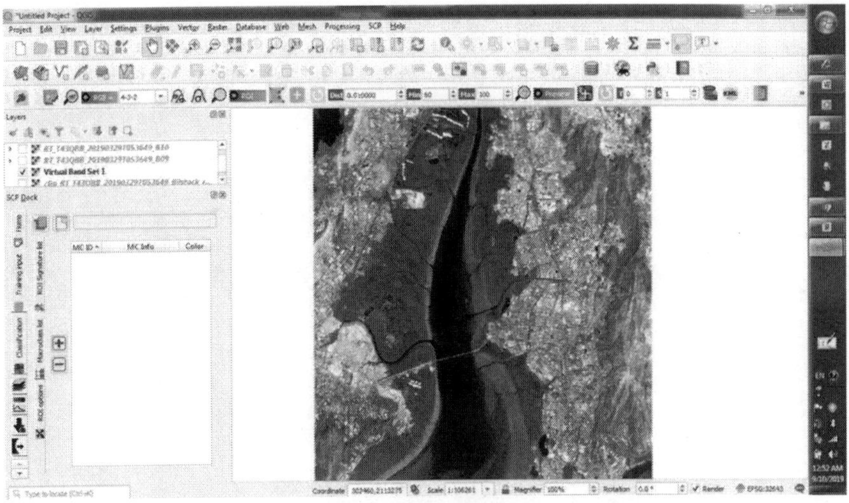

Figure 14.9a: ROI

In order to collect ROIs we need to Create the Training Input File as described in Tutorial 1:Your First Land Cover Classification (in the SCP dock click the button and define a file name). The Training input stores the ROIs and the Spectral Signature thereof.

Open the Classification dock to define the Classes and Macroclasses . In the ROI creation set MC ID = 1 and MC Info = Water; also set C ID = 1 and C Info = Lake. Nowclick to save the ROI in the Training input.

After a few seconds, the ROI is listed in the ROI Signature list and the spectral signature is calculated (because Calculate sig. was checked)

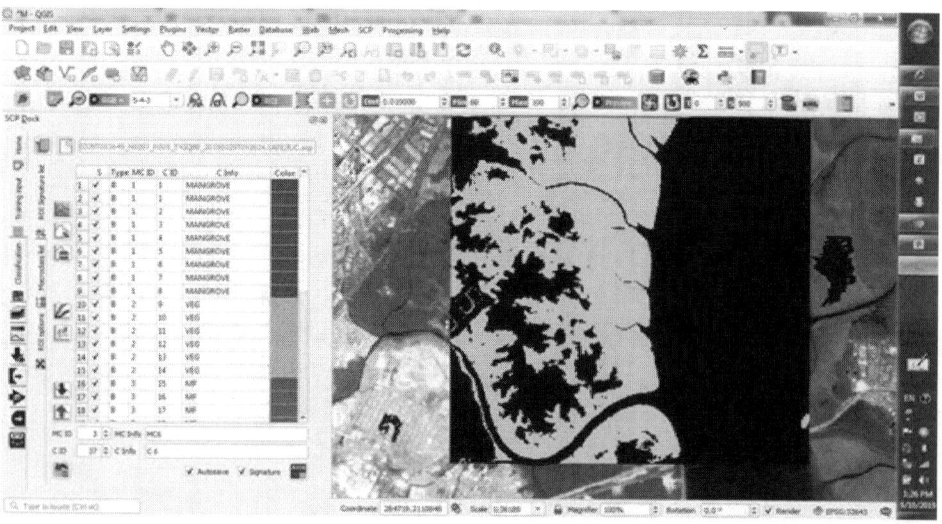

Figure 14.9b: ROI creation

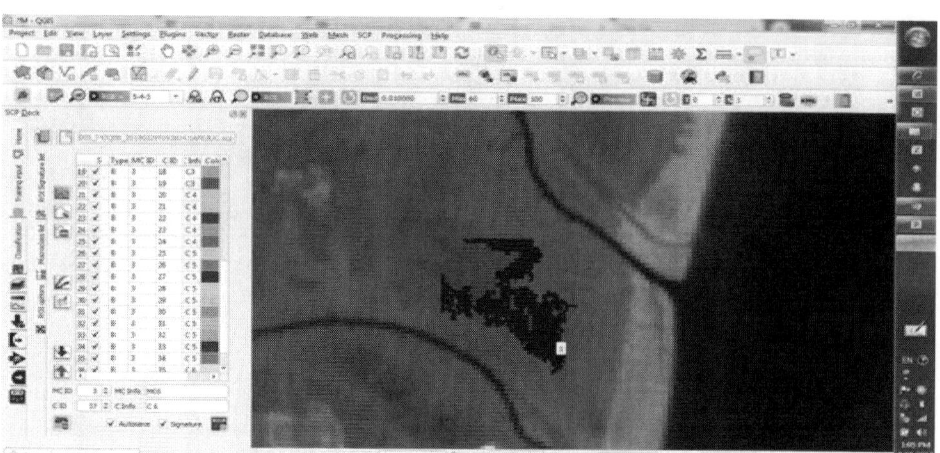

Figure 14.10: Spectral signature

5. Assess Spectral Signatures

Spectral signatures are used by **Classification Algorithms** for labelling image pixels. Different materials may have similar spectral signatures (especially considering multispectral images) such as built-up and soil. If spectral signatures used for classification are too similar, pixels could be misclassified because the algorithm is unable to discriminate correctly those signatures. Thus, it is useful to assess the **Spectral Distance** of signatures to find similar spectral signatures that must be removed. Of course, the concept of distance vary according to the algorithm used for classification. One can simply assess spectral signature similarity by displaying a signature plot. In order to display the signature plot, in the **ROI Signature list** highlight two or more spectral signatures (with click in the table), then click the button . The **Spectral Signature Plot** is displayed in a new window. Move ans zoom inside the Plot tosee if signatures are similar (i.e. very close). We can see in the following figure a signature plotof different materials.

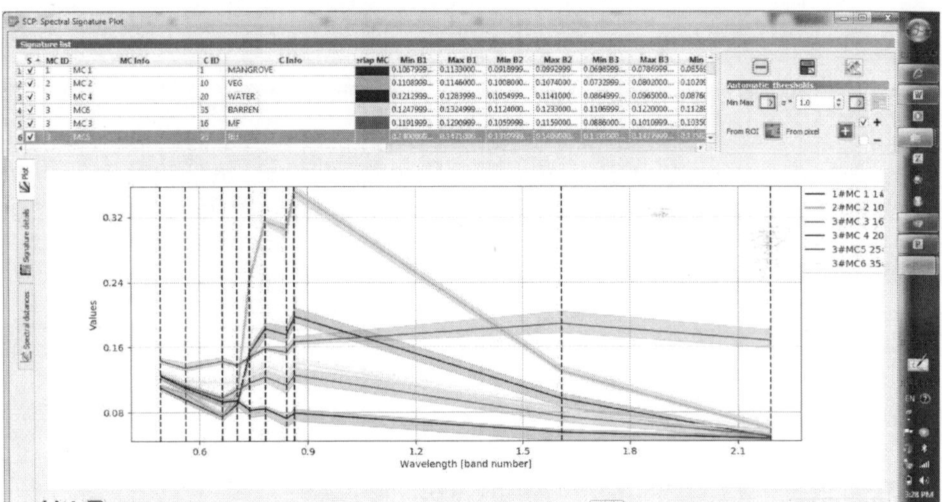

Figure 14.11: Spectral signature graph

In the plot we can see the line of each signature (with the color defined in the **ROI Signature list**), and the spectral range (minimum and maximum) of each band (i.e. the semi-transparent area colored like the signature line). The larger is the semi-transparent area of a signature, the higher is the standard deviation, and therefore the heterogeneity of pixels that composed that signature. Spectral signature values are displayed in the **Signature details**.

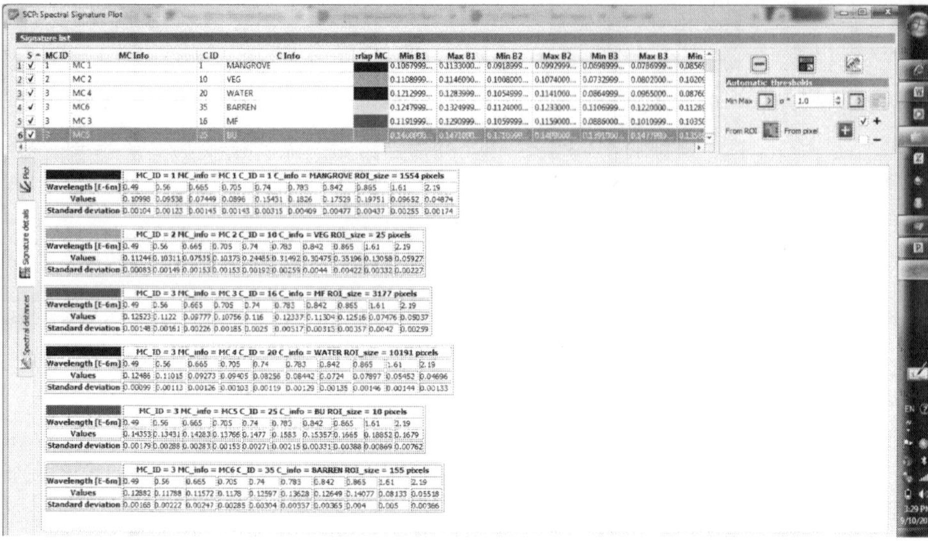

Figure 14.12: Signature details

Additionally, we can calculate the spectral distances of signatures (for more information see **Spectral Distance**). Highlight two or more spectral signatures with click in the table **Plot Signature list**, then click the button ; distances will be calculated for each pair of signatures. Now open the tab **Spectral distances**; we can notice that similarity between signatures vary according to considered algorithm.

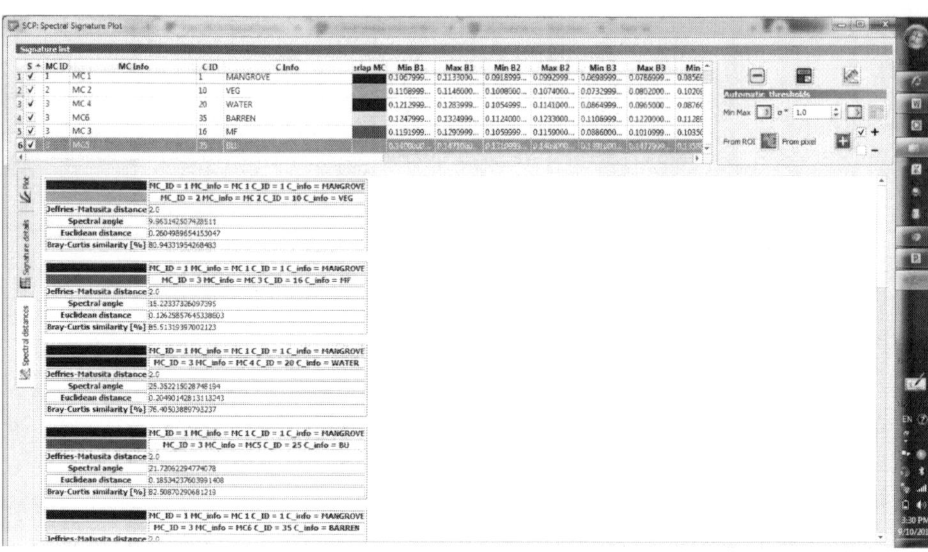

Figure 14.13: Signature distances

Create a Classification Preview

As pointed out in Tutorial 1: Your First Land Cover Classification, previews are temporary classifications that are useful for assessing the effects of spectral signatures during the ROI collection.

Set the colors of the spectral signatures in the ROI Signature list; then, in the Classification algorithm select the classification algorithm Maximum Likelihood.

In Classification preview set Size = 500; click the button and then left click a point of the image in the map. The classification preview is displayed in the map.

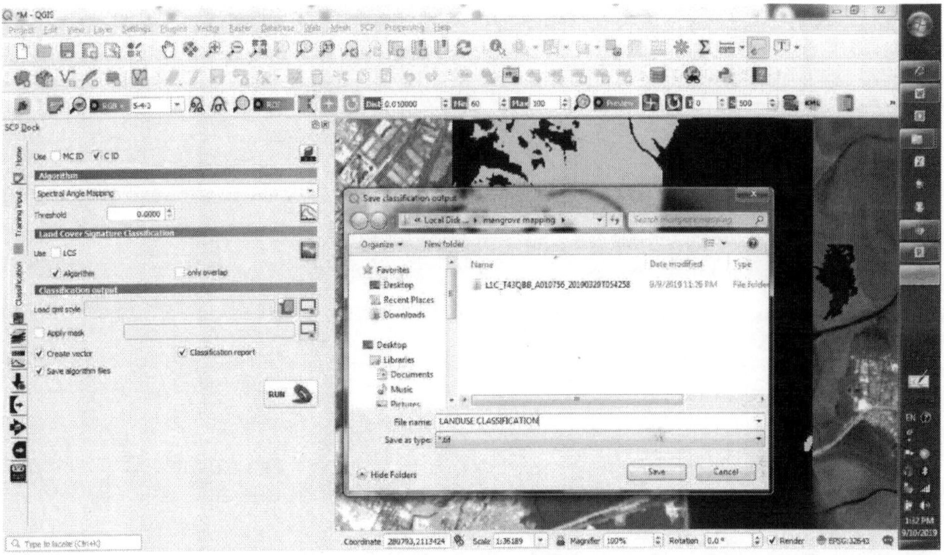

Figure 14.14: Creating the classification preview

7. Create the Classification Output

Repeat iteratively the phases Create the ROIs, Create a Classification Preview, and Assess Spectral Signatures until the classification previews show good results. To start the classification of the entire image, open the tab Classification output, click the button and define the name of the classification output.

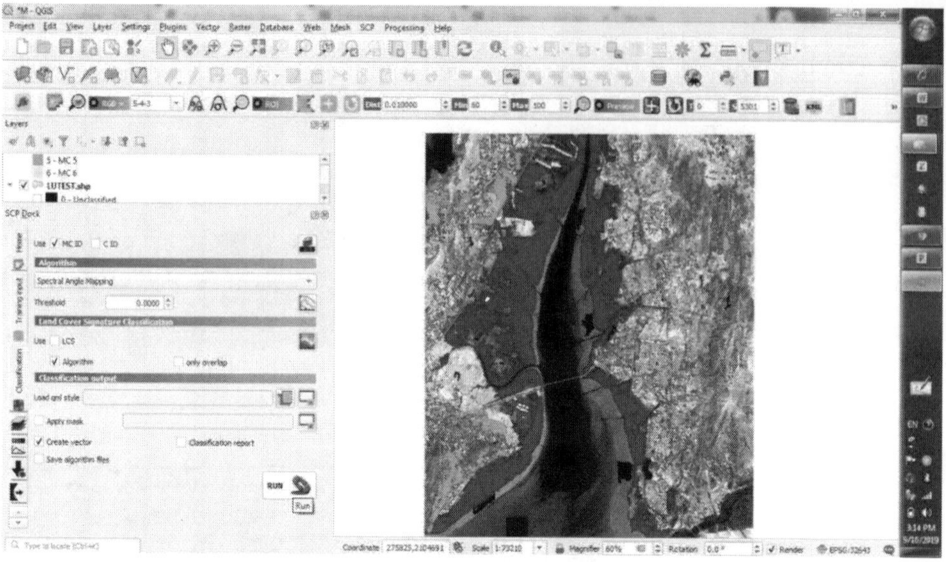

Figure 14.15: Classification output

COLOUR PLATES

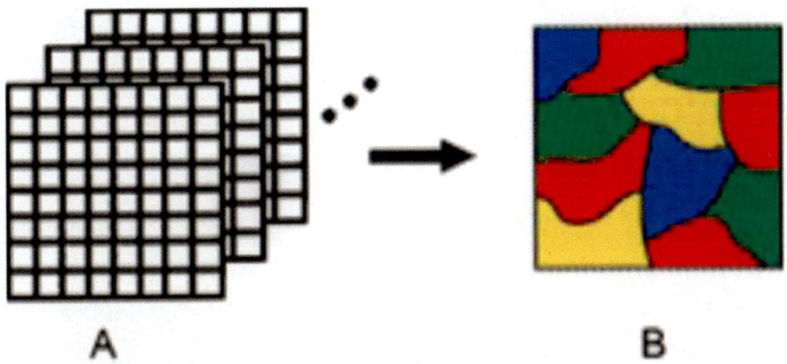

Figure 2.7: Image classification (Page 33)

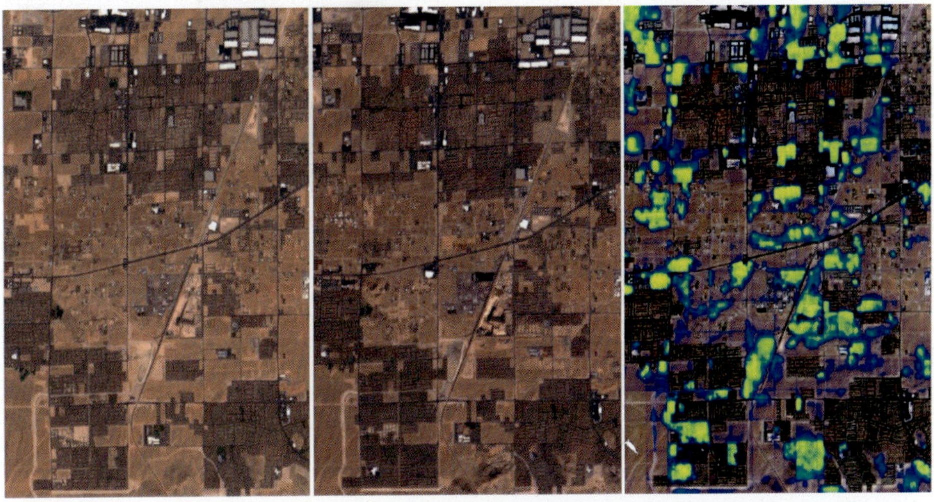

Figure 2.12: Land cover change detection (Page 40)

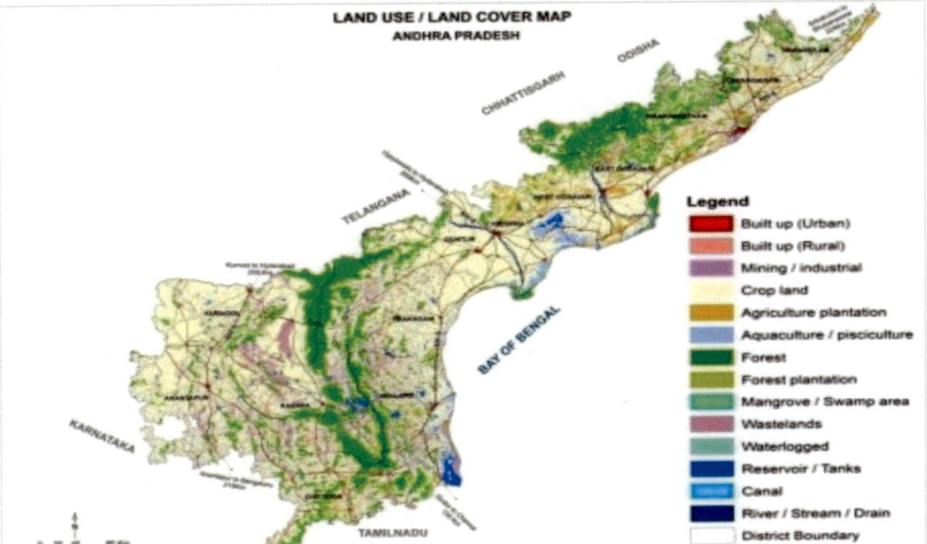

Figure 2.13: Land use land cover map of Andhra Pradesh (Page 41)

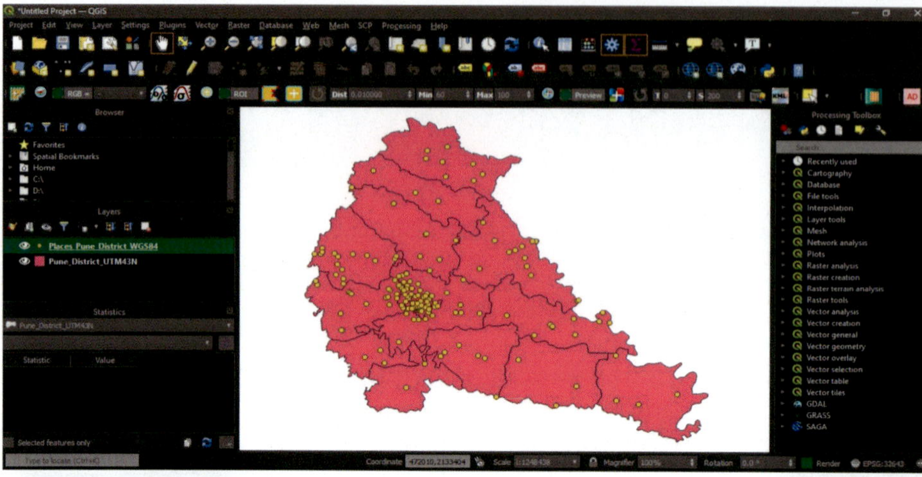

Figure 7.6: Step 6, Add another vector layer – "Places_Pune_District" (Page 124)

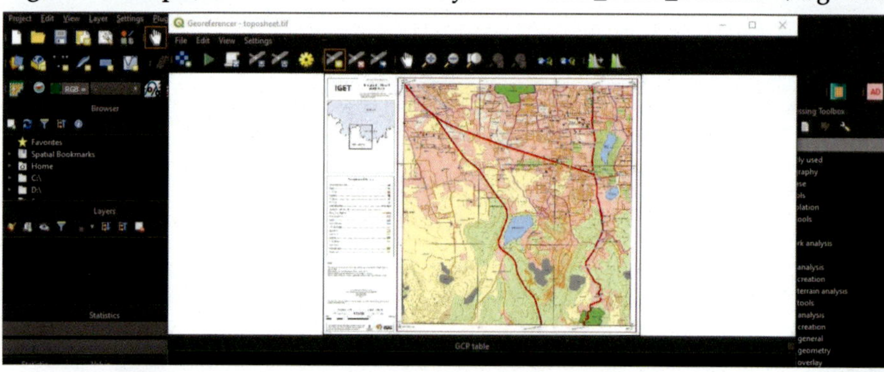

Figure 7.22: Toposheet in georeferencer window (Page 133)

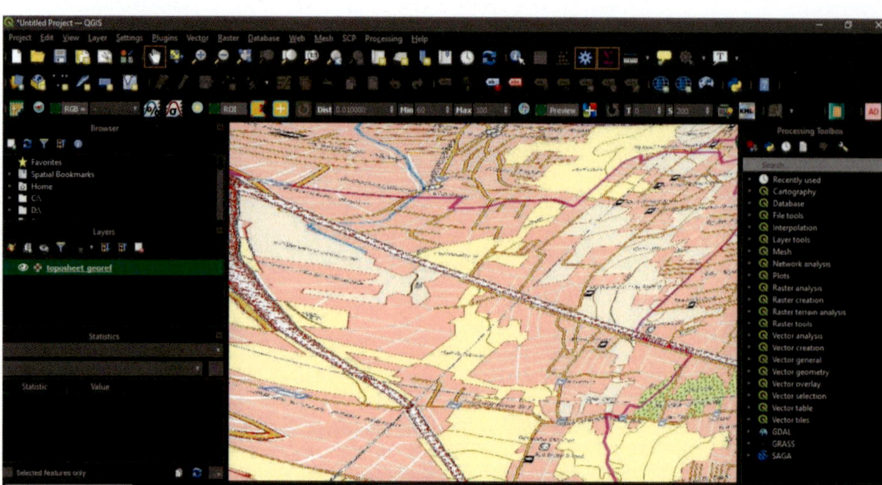

Figure 7.31: Georeferenced toposheet (Page 138)

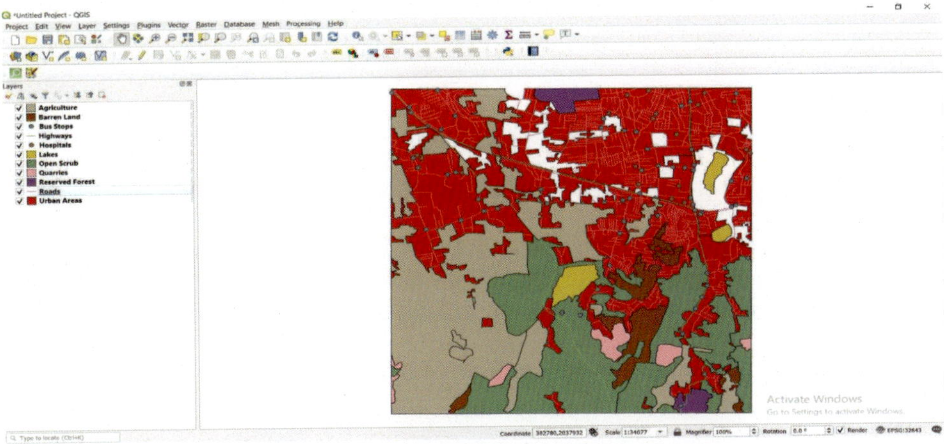

Figure 7.33: Added vector layers (Page 139)

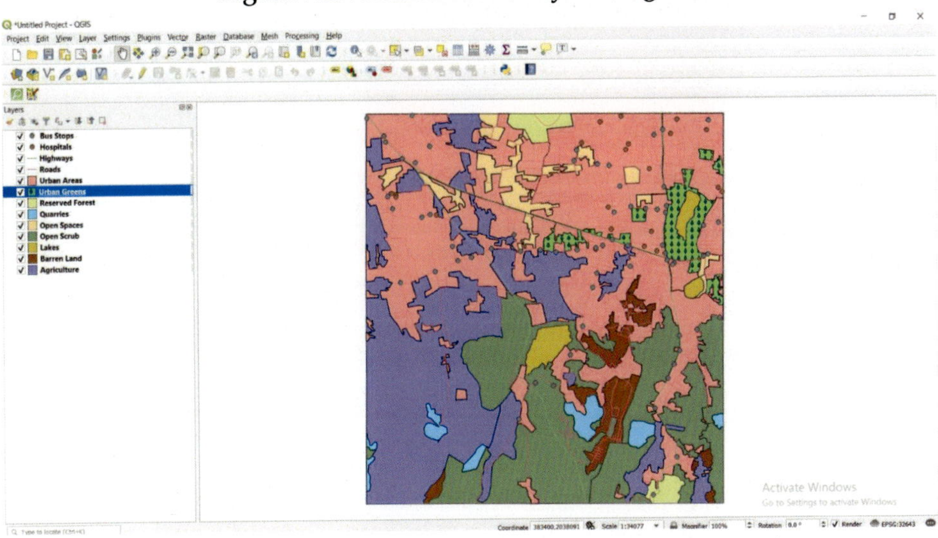

Figure 7.40: Changed pattern of the polygone (Page 143)

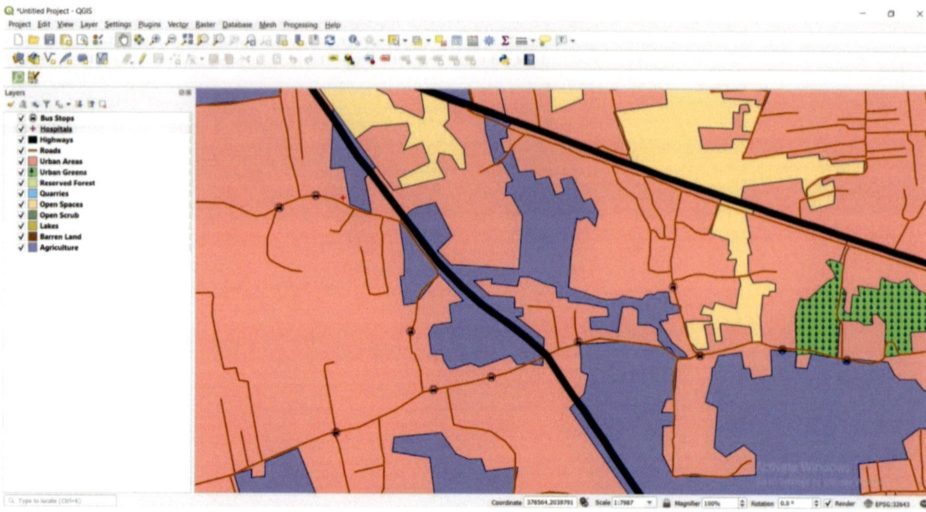

Figure 7.48: Zoom in for better visualization (Page 147)

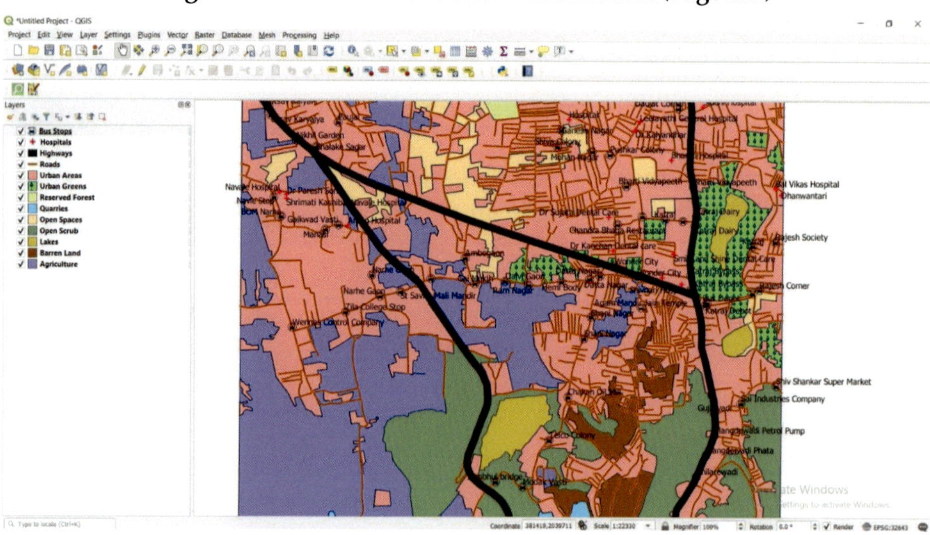

Figure 7.55: Labels of each layer (Page 150)

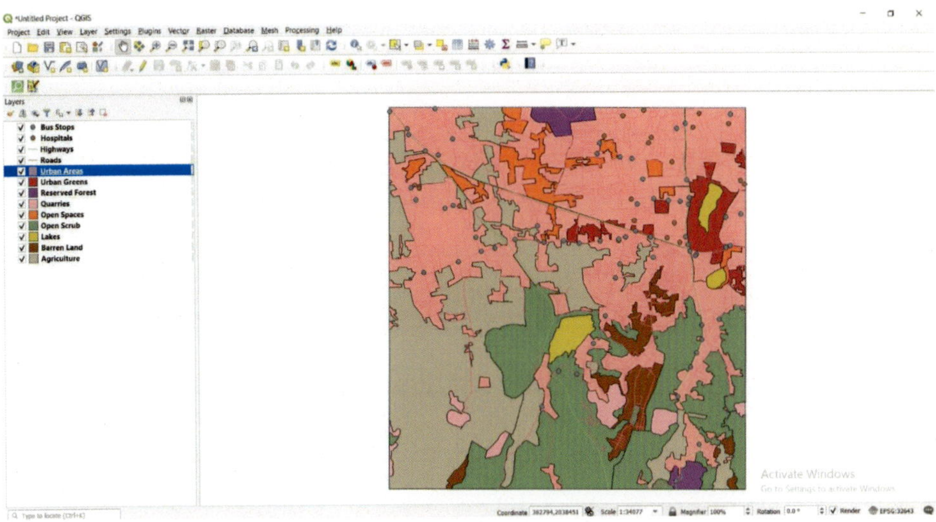

Figure 7.37: Color changed (Page 141)

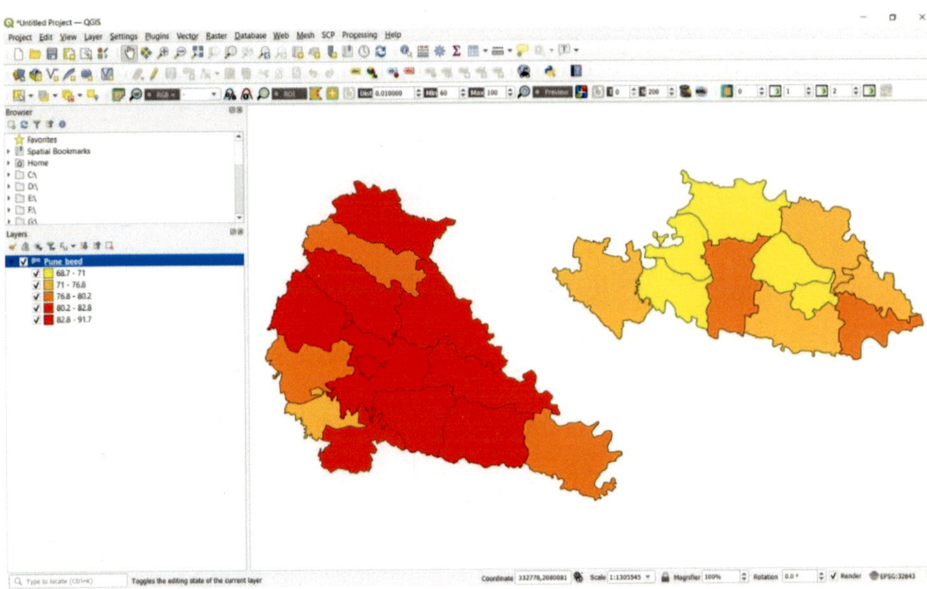

Figure 8.15: Classify and change the color (Page 864)

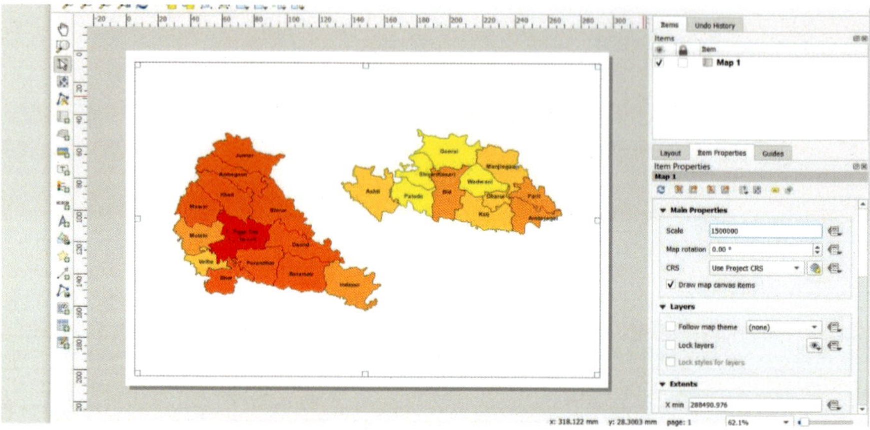

Figure 8.26: Change the scale (Page 170)

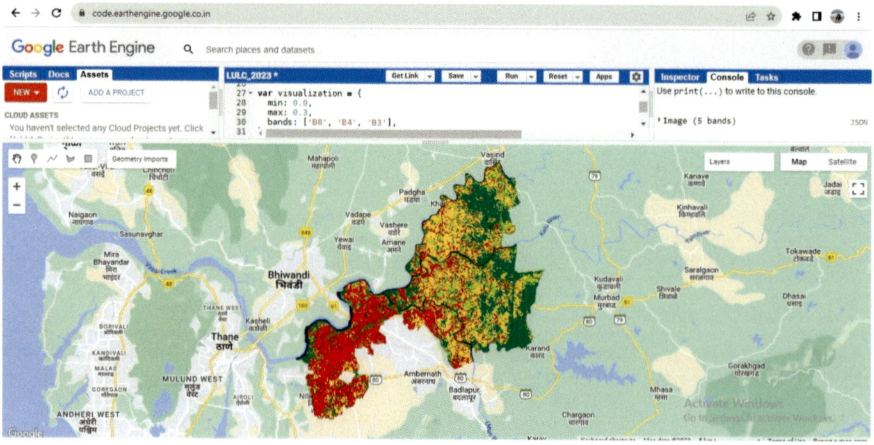

Figure 12.2: Classified image (Page 197)

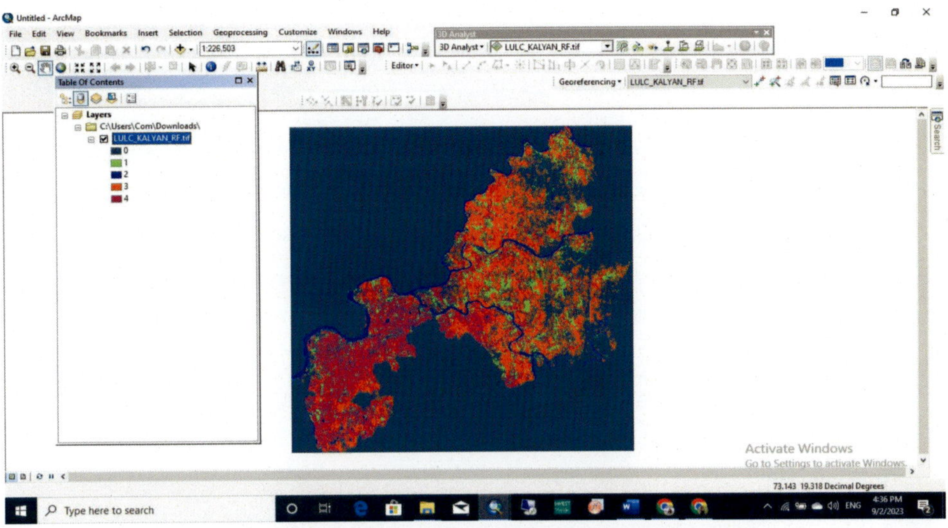

Figure 12.6: Output in the GIS software (Page 200)

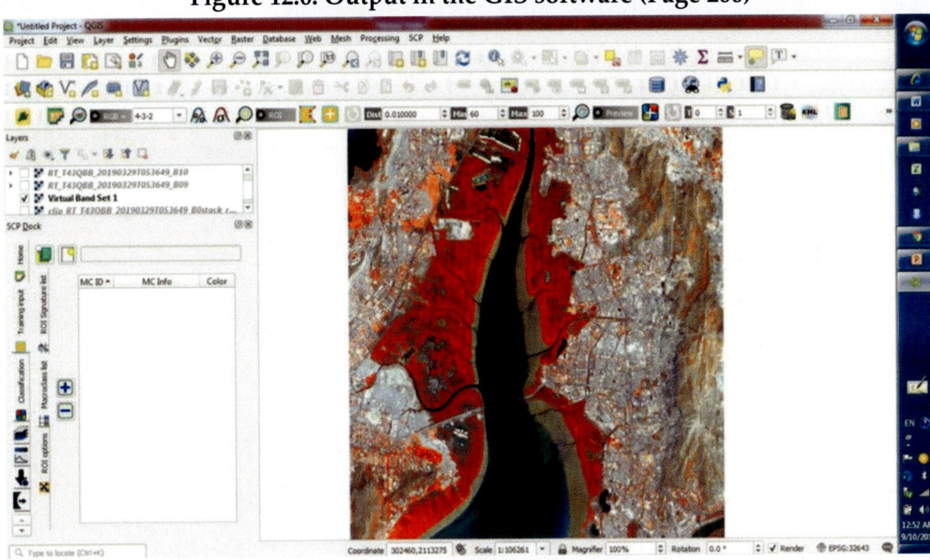

Figure 14.9: ROI (Page 221)

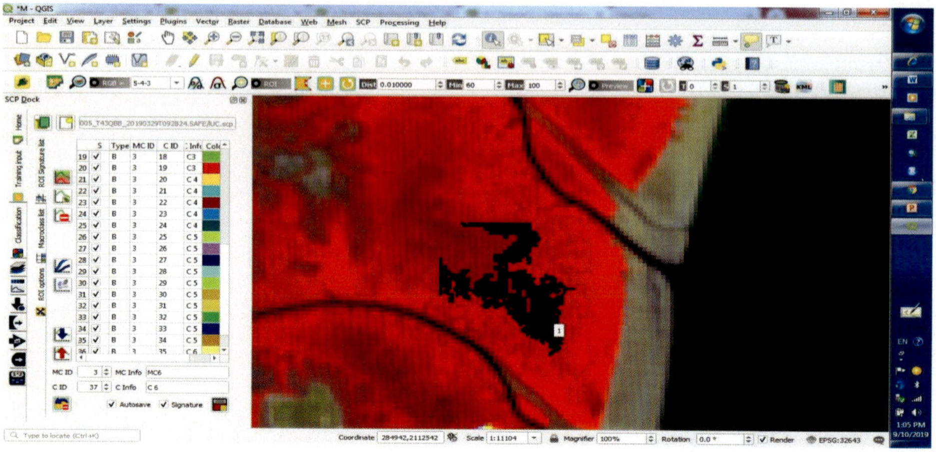

Figure 14.10: Spectral signature (Page 222)

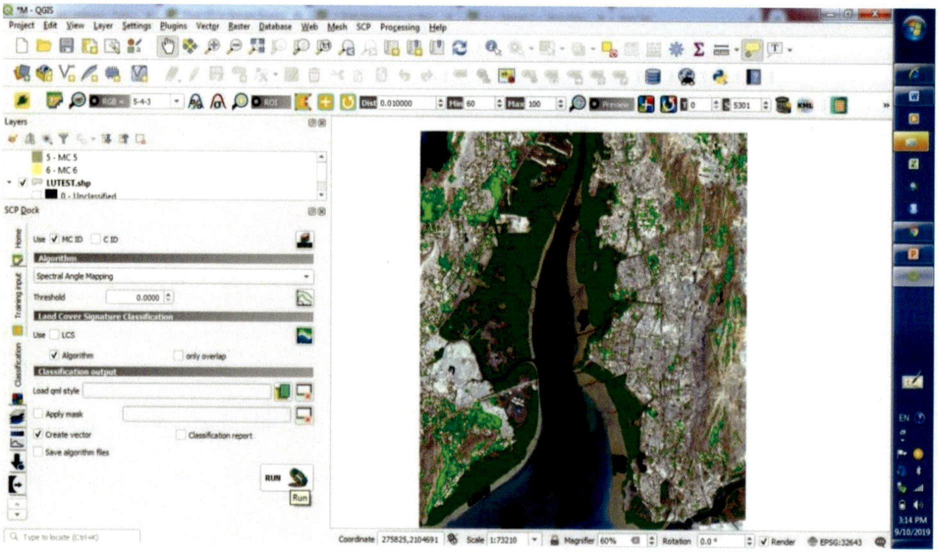

Figure 14.15: Classification output (Page 226)

INDEX